The Maintenance Costs of Aging Aircraft

Insights from Commercial Aviation

Matthew Dixon

Prepared for the United States Air Force

Approved for public release; distribution unlimited

PROJECT AIR FORCE

The research reported here was sponsored by the United States Air Force under Contract F49642-01-C-0003. Further information may be obtained from the Strategic Planning Division, Directorate of Plans, Hq USAF.

Library of Congress Cataloging-in-Publication Data

Dixon, Matthew C.
The maintenance costs of aging aircraft : insights from commercial aviation / Matthew Dixon.
p. cm.
Includes bibliographical references.
ISBN-13: 978-0-8330-3941-5 (pbk. : alk. paper)
1. Airplanes, Military—United States—Maintenance and repair. 2. United States. Air Force—Aviation supplies and stores. I. Title.

UG1243.D568 2006
358.4'183—dc22

2006028468

Published 2006 by the RAND Corporation
1776 Main Street, P.O. Box 2138, Santa Monica, CA 90407-2138
1200 South Hayes Street, Arlington, VA 22202-5050
4570 Fifth Avenue, Suite 600, Pittsburgh, PA 15213-2665
RAND URL: http://www.rand.org/
To order RAND documents or to obtain additional information, contact
Distribution Services: Telephone: (310) 451-7002;
Fax: (310) 451-6915; Email: order@rand.org

Preface

The United States Air Force is grappling with the challenge of aging fleets and when it might be optimal to replace those fleets. The RAND Corporation has worked closely with the Air Force to address these issues.

This monograph, derived from the Pardee RAND Graduate School dissertation of Air Force Captain Matthew Dixon, focuses on a specific component of the Air Force's inquiry regarding the replacement of aging fleets. In particular, it examines commercial aviation data with the goal of drawing inferences and lessons about aging aircraft that may be relevant to the Air Force. This study has methodological similarities to that of Pyles (2003), but whereas Pyles studied military aircraft, here the focus is on commercial aviation. The parameters estimated in this document might be fed into repair-replace calculations of the sort discussed in Greenfield and Persselin (2002) and Keating and Dixon (2003).

This work was sponsored by the Vice Chief of Staff, Headquarters, United States Air Force (AF/CV); Military Deputy, Office of the Assistant Secretary of the Air Force for Acquisition, Headquarters, U.S. Air Force (SAF/AQ); Deputy Chief of Staff for Air, Space, and Information Operations, Plans, and Requirements, Headquarters, U.S. Air Force (AF/A3/5); and Deputy Chief of Staff for Logistics, Installations, and Mission Support, Headquarters, U.S. Air Force (AF/A4/7). It was performed as part of a fiscal year 2005 project entitled "When to Recapitalize." This monograph should be of interest to Air Force and other Department of Defense acquisition, financial, and maintenance personnel.

RAND Project AIR FORCE

RAND Project AIR FORCE (PAF), a division of the RAND Corporation, is the U.S. Air Force's federally funded research and development center for studies and analyses. PAF provides the Air Force with independent analysis of policy alternatives affecting the development, employment, combat readiness, and support of current and future aerospace forces. Research is conducted in four programs: Aerospace Force Development; Manpower, Personnel, and Training; Resource Management; and Strategy and Doctrine. The research reported here was conducted within the PAF-wide program.

Additional information about PAF is available on our Web site at http://www.rand.org/paf.

Contents

Figures

Tables

Summary

This monograph examines "aging effects"—i.e., how commercial aircraft maintenance costs change as aircraft grow older. Although commercial aircraft clearly differ from military aircraft, commercial aviation aging-effect estimates might help the Air Force to project how its maintenance costs will change over time.

Literature Discussion and Prior Work

There is a large body of literature on aging aircraft, much of which focuses on military aviation. Recent studies have generally found positive aging effects (costs rising with age), although the estimated magnitude of the effects has varied considerably (see pp. 5–13).

Boeing's 2004 analysis of commercial aviation aging effects (Boeing, 2004a) is the most direct intellectual forerunner to this current study. Boeing computed a "maturity curve" for airframe maintenance costs. Boeing found airlines' airframe maintenance costs increase as aircraft come off warranty, then enter a stable "mature" period after the first D check[1] (depot-level heavy maintenance), and then resume rising after about 10–14 years of service and the second D check (see pp. 13–15). Of course, the observed jump in aircraft maintenance costs as aircraft come off warranty does not represent an increase in maintenance as much as a transfer of maintenance cost responsibility from the aircraft's manufacturer to its owner.

[1] A *D check* is a complete structural check and restoration.

Commercial Aviation Maintenance Data

Form 41 data are reports that U.S. commercial airlines are required to file with the Department of Transportation (DoT) indicating their maintenance costs and flying hours. RAND gathered Form 41 data from the DoT on maintenance costs going back to the 1960s. Separately, RAND obtained data on airlines' average fleet ages by calendar year.

The estimation strategy was to run a log linear regression with the natural logarithm of maintenance cost per flying hour as the dependent variable and various independent variables including average fleet age. The coefficient on the age variable in such a regression would estimate the age effect, i.e., how maintenance costs typically change as aircraft age, other things being equal.

Results

The RAND study team ran three separate log linear regressions, computing age effects for aircraft 0–6 years old, 6–12 years old, and more than 12 years old. Figure S.1 depicts the results (with the total maintenance costs per flight hour for a six-year old aircraft normalized to 1.0).

This study found that young aircraft have considerable age effects (an estimated 17.6 percent annual rate of increase in maintenance cost per flying hour, with a standard error on that estimate of 1.8 percent). This age effect reflects aircraft coming off warranty, which increases airline maintenance costs.

For mature aircraft, ages 6–12, a 3.5 percent annual age effect was found, with a standard error of 0.8 percent.

Most intriguingly, an age effect of 0.7 percent, not statistically significantly different from zero, was computed for aircraft over 12 years of age (see pp. 27–28).

One reason that these findings differ from Boeing's maturity curve is that RAND analyzed total maintenance costs, including engine and

Figure S.1
Age Effects Estimated with Form 41 Data

RAND *MG486-S.1*

overhead costs, not simply airframe maintenance costs. Airframe-maintenance cost growth shows a more convex growth pattern than Figure S.1's depiction of total maintenance cost growth. Engine maintenance costs, by contrast, seem to remain very flat as aircraft age (after an initial jump in the first years of operation). Airframe maintenance costs are only about a third of total maintenance costs in the data analyzed (see pp. 29–31).

RAND experimented with other regression specifications, e.g., airline-specific dichotomous (dummy) variables and endogenous selection of age breaks. None of these alternative specifications provided meaningfully different findings (see pp. 32–35).

Potential Bias in Estimated Age Effect

The study team was concerned that airlines were prematurely retiring "poorly aging" fleets and that such early retirements caused Figure S.1 to be artificially concave.

The study team analyzed 21 fleets that were retired before an average age of 20 years. The team did not find evidence that the fleets had unusual aging effects. It was found, however, that these early-retired fleets were unusually expensive in the first 12 years of their lives. Cost problems may have encouraged airlines to retire these fleets, but there was no evidence that those problems were worsening unusually rapidly (see pp. 37–40).

RAND did not find that fleet-level retirement selection bias causes Figure S.1's concavity.

Conclusions

If one believes that commercial aviation experience is germane to the Air Force, this study suggests that total aircraft maintenance costs may plateau, at least for certain aircraft ages. Pessimism about the future trajectory of total maintenance costs may not always be correct. RAND also found different cost patterns for different types of aircraft maintenance, e.g., airframe maintenance versus engine maintenance.

Acknowledgments

The author is grateful to the United States Air Force, the Air Force Institute of Technology, and Colonels Jerry Diaz and Daniel Litwhiler of the United States Air Force Academy for providing the opportunity to study at the Pardee RAND Graduate School (PRGS). Natalie Crawford and Michael Kennedy of RAND Project AIR FORCE have been very helpful and supportive.

Thank you to my committee members, Bart Bennett, Ed Keating, and Greg Ridgeway. Special thanks are due Ed Keating for his mentorship, patience, and coaching during my three years at PRGS. Much of my success at PRGS is due to his support and diligence. The chairman, Bart Bennett, has also been a great help throughout my studies at PRGS, and he provides invaluable direction and encouragement for all the students. Greg Ridgeway has provided important help, feedback, and encouragement throughout this dissertation process.

The author received helpful PAF internal reviews of this document from Rob Leonard and Raj Raman. Cynthia Cook provided careful quality assurance oversight. Jane Siegel helped to prepare and format a previous version of this document. Nancy DelFavero edited the final document. Bob Roll and Laura Baldwin provided useful direction; Baldwin originally proposed this dissertation topic. Nancy Moore provided insights on commercial aviation maintenance practices. Rodger Madison provided computing assistance. Greg Hildebrandt and Ray Pyles assisted with the literature review in Chapter Two.

Thank you to Dan Abraham of Boeing for valuable data, responses to myriad questions, and insight into the commercial aviation world. Professor Douglas C. Montgomery of Arizona State University provided helpful comments on an earlier draft. An earlier version of this research was briefed at Boeing in Renton, Washington, on June 29, 2005; the comments and suggestions of seminar participants were most appreciated.

Many of the data analyzed in this report were originally gathered for RAND by Jean Gebman, Priscilla Schlegel, and Elaine Wagner.

Of course, the author alone is responsible for any errors that remain in this report.

Acronyms

ALC	Air Logistics Center
CBO	Congressional Budget Office
CNA	Center for Naval Analyses
COTS	commercial, off-the-shelf
df	degrees of freedom
DoD	Department of Defense
DoT	Department of Transportation
IPT	Integrated Product Team
MS	mean squared
MSE	mean squared errors
MTBF	Mean Time Between Failures
NAMO	Naval Aviation Maintenance Office
OC-ALC	Oklahoma City ALC
O&S	Operating and support
PAF	RAND Project AIR FORCE
PDM	Programmed Depot Maintenance
PRGS	Pardee RAND Graduate School

REMIS	Reliability and Maintainability Information System
SS	sum of squares
TMS	Type Model Series (U.S. Navy terminology)
USAF	United States Air Force

CHAPTER ONE
Introduction

The United States Air Force is interested in estimating how maintenance costs associated with its various aircraft will change over time. The Air Force is also interested in how maintenance costs might evolve for new aircraft not yet in its inventory. Future maintenance cost projections are important for budgeting purposes, but they are also central to optimal aircraft replacement calculations of the sort done by Greenfield and Persselin (2002) and Keating and Dixon (2003). If an existing aircraft's maintenance costs grow more quickly, its optimal replacement date will move forward. Conversely, if a replacement aircraft is projected to have rapidly escalating maintenance costs, the Air Force may wish to hold on to an existing aircraft longer.

Pyles (2003) is a fairly recent, and quite exhaustive, analysis of "age effects" (i.e., how maintenance costs change as aircraft grow older) in military aircraft. (The literature review in Chapter Two has further discussion of the Pyles study and other analyses of military aircraft.) This report complements the literature on aging military aircraft by focusing instead on commercial aviation.

There are, obviously, important differences between commercial and military aviation. Commercial aircraft are operated many more hours per day—a commercial aircraft might have ten times as many lifetime flying hours as a military aircraft of similar age.

Perhaps as a result of fewer flight hours per year, the Air Force is currently operating some aircraft (e.g., the B-52, the KC-135) at ages not seen in U.S. commercial aviation. As discussed in Chapter Three, commercial aircraft are generally disposed of by U.S. airlines by around age 25. Hence, the analysis in this document is not informative as to what might happen to maintenance costs of the Air Force's oldest aircraft.

Of course, commercial aviation is not intended to operate in the hostile conditions of combat. For instance, damage from anti-aircraft weapons or super-normal gravitational forces should not be observed in commercial aviation.

Why, then, might commercial aviation be of interest to the Air Force? There are several possible motivators for analyzing commercial aircraft costs in order to gain insights on military aircraft maintenance costs, although the reader must ultimately decide on the relevance of this study.

First, the Air Force owns and/or is considering purchasing aircraft that have commercial analogs. The Air Force's executive transport aircraft are essentially commercial-off-the-shelf (COTS) except with military communications (e.g., identification, friend or foe) equipment installed. More importantly, the Air Force's cargo and tanker aircraft are similar to commercial passenger aircraft. The Air Force, for instance, is currently considering acquiring a tanker variant of an existing Airbus and/or Boeing commercial passenger airliner. While such a commercially derived tanker would not be equivalent to its passenger cousin, it is reasonable to think its maintenance issues could be analogous.

Second, the RAND study team hypothesized that some commercial aviation aging effects may be similar to those of the Air Force, notwithstanding major differences in usage. At the risk of gross oversimplification, there are two basic causes of maintenance costs. The more intuitive cause is usage: Every time an aircraft takes off or lands or flies for an hour, a certain amount of wear and tear occurs that requires maintenance. The less intuitive cause of maintenance is time itself: Destructive processes, such as corrosion or seals drying out, occur irrespective of whether an aircraft is flying. Commercial experience is especially relevant to the Air Force to the extent that calendar-age-related maintenance costs are important.

Third, commercial aviation maintenance cost data have been collected that may prove to be more comprehensive and more detailed than military maintenance cost data. Pyles (2003), for instance, observed that

> the Air Force has no comprehensive system for historical maintenance and material consumption data. Some historical data exist only as hard-copy records kept in office file cabinets or in old reports archived sporadically.

Chapter Three discusses the commercial aviation data gathered by the Department of Transportation (DoT) that were used in this study. Although all data sets have shortcomings, these DoT data provide a 1965–2003 annual time series that goes far beyond the duration of most military maintenance data sets.

The remainder of this monograph is organized as follows: Chapter Two presents a literature review on aging aircraft. Chapter Three commences with a simplified overview of how commercial aircraft are maintained and then discusses the DoT commercial aviation maintenance data on which this document is based. Chapter Four presents the results of the analysis. It presents estimates of how commercial aircraft maintenance costs typically change as commercial aircraft grow older. Chapter Five discusses a prospective bias in the estimation. Specifically, the concern is that commercial airlines might be prematurely retiring "poorly aging" fleets—an option probably not available to the Air Force. Fortunately, evidence of such an effect was not found. Chapter Six provides the conclusions, and a technical appendix provides detailed results of the estimations.

CHAPTER TWO

Literature and Prior Work on Aging Aircraft

This chapter discusses the literature and prior research relating to aging-aircraft issues. While research undertaken in the 1960s did not consistently find maintenance costs increasing as aircraft aged, more recent studies have generally found an aging effect. Table 2.1 summarizes previous studies in chronological order. The "age effect" column has a "+" in it if the study found a positive age effect—i.e., real (inflation-adjusted) maintenance costs grew as aircraft aged. The studies looked at multiple models and explanatory variables. Some combinations yielded no age effect, while others did. A "No" in the age effect column indicates that there was no age effect worth reporting in the analysis. None of the authors of these studies reported negative age effects. The next section covers each of the studies individually.

Chronology of Prior Studies

Since the advent of aviation maintenance, those responsible for maintaining aircraft have been concerned not only with the current cost of maintenance but also the future cost. The Air Force is no exception. Studies going back to the 1960s demonstrate the Air Force's historical concern over the expected future cost of its fleet maintenance.

Kamins (1970) Found Lack of Age Effect

In a RAND study published in 1970, Kamins cited ten different analyses that attempt to illustrate the effect of age on maintenance cost. He briefly critiqued three studies that show a positive age effect but argued

Table 2.1
Summary of Literature Related to Aging Aircraft

Authors	Date	Age Effect	Data Level	Sector	Primary Dependent Variable	Data Type
Kamins (RAND)	1970	No	Multiple	Air Force and Commercial	Multiple	Cross-sectional and panel (two-period)
Hildebrandt and Sze (RAND)	1990	+	Aircraft	Air Force	Operating and support (O&S)/ Aircraft	Panel
Johnson (Naval Aviation Maintenance Office [NAMO])	1993	+	Aircraft	Navy	Mean time between failures (MTBF)	Cross-sectional
Stoll and Davis (NAMO)	1993	+	Multiple	Navy	Multiple	Cross-sectional and panel
Ramsey (Oklahoma City Air Logistics Center [OC-ALC]), French and Sperry (Boeing)	1998	+	Multiple	Air Force and Commercial	Programmed depot maintenance (PDM) man-hours	Panel
Francis and Shaw (Center for Naval Analyses [CNA])	2000	+	Aircraft	Navy	Maintenance man-hours	Panel
Kiley (Congressional Budget Office [CBO])	2001	+	Anecdotal	Air Force	Operations cost/ Flight hours	Panel

Table 2.1—Continued

Authors	Date	Age Effect	Data Level	Sector	Primary Dependent Variable	Data Type
Jondrow et al. (CNA Corporation)	2002	+	Aircraft	Navy	Repairs/Flight hours	Panel
Pyles (RAND)	2003	+	Aircraft	Air Force	Workloads and material consumption	Cross-section and Panel
Boeing	2004	+	Fleet	Commercial	Cost/Flight hours	Panel

that the studies are insufficient, primarily because the data were cross-sectional, the data points were few, and the representation of aircraft of various ages was skewed and over-represented by older aircraft. In the early studies, the two aircraft of interest were the B-52 and the KC-135A.

Kamins then moved to seven studies he said prove that there is no age effect. In fact, some of the studies seemed to demonstrate that aircraft actually become more reliable as they age. One study used accidents as the dependent variable, with the argument that accident rates decreased as aircraft got older, thus demonstrating a negative age effect. A second study summarized findings from United Airlines and Pan American Airlines that stated that due to process improvements in maintenance, maintenance requirements actually decreased as aircraft aged.

The studies that were used to justify the lack of age effect have a small number of observations. Extrapolations of any results were nearly impossible. These studies were completed while aviation was still in its youth and when aircraft were retired because of technological advances and not because of maintenance costs.

Kamins demonstrated that aging aircraft are not unique to the military. The airlines and aircraft manufacturers are equally, if not more, concerned than the Air Force with growth in the cost of maintenance. However, most of the available literature regarding age effects focuses on military fleets.

Hildebrandt and Sze (1990) Found Positive Age Effects

Hildebrandt and Sze (1990) developed several O&S cost-estimating relationships in which models were estimated to determine the effect of specified explanatory variables on different aggregations of O&S cost. Aircraft mission design age was included in their analysis as one of the explanatory variables. They emphasized the explanatory power of a total O&S cost model in which flyaway cost is a proxy for both aircraft mission type and the year an aircraft entered the inventory. They also examined specifications that attenuated the fact that flying hours are used, in many cases, to allocate costs to an aircraft mission design series. For a depot maintenance model, they estimated an aging effect of about 2.0 percent per year of aircraft design age. For the aircraft overhaul subcategory, they found that a one-year increase in aircraft mission design age increases costs by about 3.1 percent.

While Hildebrandt and Sze estimated the allocation of the funds to specific maintenance costs, the commercial data used in their report contain the actual dollars spent annually on labor, materials, and overhead for maintenance for a specific fleet.

Johnson (1993) and Stoll and Davis (1993) Found Evidence of Larger Age Effects[1]

The Naval Aviation and Maintenance Office (Johnson, 1993) found significant age effects on total maintenance workloads in naval aircraft over a 13-year period. Also in 1993, Stoll and Davis found smaller naval aircraft age effects in on-equipment[2] workloads over approximately the same period of time.

Ramsey, French, and Sperry (1998) Used Commercial Data to Estimate KC-135 Age Effects

The Oklahoma City Air Logistics Center led a KC-135 Cost of Ownership Integrated Product Team (IPT) study (Ramsey, French, and Sperry, 1998). The purpose of the study was to develop aging-aircraft maintenance cost trends for the KC-135 based on a review of historical commercial and military data. Ramsey, French, and Sperry used military data from the Air Force combined with 12 years of commercial panel data from the DoT. They used aircraft types similar in structure, size, and composition to the KC-135. They reported varying annual airframe maintenance cost growth rates for various commercial aircraft, e.g., 3.5 percent for DC-9s, 9 percent for DC-10s.

---

[1] Pyles (2003) provides a more lengthy discussion of these studies.

[2] *On-equipment maintenance* refers to maintenance activities done by maintenance personnel directly on an aircraft. For instance, changing a tire would be on-equipment maintenance. Its antonym is *off-equipment maintenance,* in which a broken part is removed from an aircraft, fixed in a back shop, and then later returned to the aircraft.

Francis and Shaw (2000) and Jondrow et al. (2002) Demonstrated Positive Age Effects for Navy Aircraft

The Center for Naval Analyses analyzed the Navy's F/A-18 Hornets. Francis and Shaw (2000) of the CNA used two different datasets to gain information about F/A-18 maintenance costs. Both datasets have information on the individual tail numbers. The first dataset contains ten years (1990–1999) worth of data about the usage and maintenance of every tail number of the F/A-18s in inventory. This information includes aircraft age, squadron manning numbers, maintenance time, deployment status, flight hours, and sorties. Their regression model used the log of maintenance man-hours as the dependent variable and several independent variables including number of flight hours, deployment status (e.g., whether the aircraft was deployed during the month in question), personnel variables, and age. They found a significant age effect. The age effect was 6.5 percent to 8.9 percent per calendar year of age. Additionally, they found that the flight hours and deployment status were significant indicators of the man-hours required for maintenance.

The second dataset contained information about every F/A-18 sortie flown in one month along with records of the surrounding maintenance activities. Francis and Shaw employed a probit model to estimate the probability that an F/A-18 would require unscheduled maintenance after a sortie. The independent variables were aircraft age, length of time since last depot-level maintenance, and an indicator for whether the sortie was carrier-based. They estimated that a one-year gain in age significantly increased (by 0.8 percent) the probability of unscheduled maintenance and a carrier-based sortie significantly increased (by 3.5 percent) the probability of needing unscheduled maintenance.

Another CNA study (Jondrow et al., 2002) found age effects for all types of Navy aircraft. Jondrow et al. used a log-linear model with parameters estimated with weighted least squares. The independent variables used were the annual hours flown, the percentage change in average age of a Type Model Series (TMS) (e.g., F-14A), and a categorization of the type of aircraft (carrier-based fixed wing, land-based fixed wing, or rotary wing). The dependent variable is the number of repairs per flight hour. Jondrow et al.'s goal was to help the Navy understand the effective cost of a new aircraft so that the Navy can make informed repair-versus-replace decisions. At the mean aircraft age in the dataset, they found repair-per-flight-hour age effects of 1.9 percent, 1.7 percent, and 7.9 percent for the land-based aircraft, rotary-wing aircraft, and carrier-based aircraft, respectively.

Jondrow et al. also found that some aircraft become significantly less expensive to maintain as they near retirement (the end of their service life). Readiness (the aircraft mission-capable rate) generally declines as aircraft age, but they found that as the F-14 and A-6 neared retirement, their readiness increased. Selective decommissioning is a cited reason for increased readiness near retirement. Another cited reason is that spare parts and maintainers do not drop proportional to the number of aircraft retired (Jondrow et al., 2002, slide 31).

Kiley (2001) Found Lower Aircraft Age Effects

In a Congressional Budget Office study, Kiley (2001) examined the age effects on all military equipment, including aircraft. The purpose of the study was to understand the rise in the military's O&S expenditures and discuss prior literature about the effects of age on O&S expenditures, which includes maintenance. Kiley did no new analysis with raw data. However, as stated in the report, "Those studies typically found that the costs of operating and maintaining aircraft increase by 1 to 3 percent with every additional year of age after adjusting for inflation" (Kiley, 2001).

Pyles (2003) Found Specific Age Effects on Workloads and Material Consumption

Pyles (2003) is the most comprehensive study of age effects on Air Force aircraft to date. This RAND study estimated multiple models for calculating how Air Force maintenance requirements change over time. Specifically, Pyles studied how aircraft age relates to maintenance and modification workloads and to material consumption. He used two conceptual models, looking first at the material consumption and workload for maintenance and then at modifications. Both models allow for varying effects at different aircraft ages, and Pyles took considerable effort to distinguish the actual age effect from other factors.

He experimented with both linear and logarithmic dependent variable specifications.

Pyles used regression analysis to address several questions about age effects, including questions on how a fleet ages, if and how platforms age differently, the future prospects for cost and workload growth, and the age effect at different ages. He analyzed trends at many different levels including at the on-equipment, off-equipment, depot, and engine levels.

Pyles found several statistically significant results in the data. He found specific age-related growths of maintenance conditional on the age of the aircraft and on the "fly-away" costs (how much the aircraft costs new, which is a measure of aircraft complexity). Furthermore, he also found that, in general, maintenance requirements increase as aircraft age, and more-expensive aircraft generally experience higher growth rates. He estimated that maintenance costs grew through age 40, and "honeymoon" periods and "infantile" failures were both present, skewing the long-term growth-rate estimates. (Infantile periods caused the growth rates to be underestimated, and honeymoon periods caused the rates to be overestimated.) He also found that material consumption decelerated as the aircraft aged.

Pyles also found that the PDM workloads grew steadily over the first 40 years of aircraft operation.

Pyles looked at many models while searching for a relationship between age and proxies for maintenance costs. In some of these models, he found effects, and in others he did not. He found that growth was not uniform across fleets, flying hours, or flyaway costs.

For example, a $30 million fighter flying 300 hours per year would require 60 additional on-equipment, base maintenance man-hours each year, while a $100 million cargo aircraft flying 500 hours per year would require 330 additional maintenance man-hours each year. Other workloads (off-equipment, engine depot maintenance, PDM, depot-level reparable repair) have different life-cycle patterns, but they also generally grow after the first few years of operations.

In summary, Pyles found a positive age effect on maintenance requirements for nearly all activities. The number of man-hours required to perform the same task increased over time. Additionally, he found that more-complex aircraft's maintenance requirements increase at a faster rate than do those of simpler aircraft.

The research questions presented in Pyles' study are similar to the research questions this report is addressing. This study builds on Pyles' study in two ways. First, it uses a completely different dataset to test similar hypotheses. Second, it measures maintenance costs in real dollars (cost per flight hour).

Boeing (2004) Provided Maturity Curves for Cost-Comparison Purposes

Few publicly available studies show the maintenance-cost age effects of commercial aircraft.

Boeing, not surprisingly, is very interested in the effect of aging on maintenance. Boeing benefits from supporting its current customers and its potential purchasers of new and used aircraft. Understanding the implications of age will help users of Boeing's products to understand the true cost of ownership. In Boeing's ongoing work on aging aircraft, the primary research objective is to understand how operating aircraft beyond their design service life will affect maintenance (Boeing, 2004a).

Boeing separates an aircraft's life into three stages: the "newness" period, the "mature" period, and the "aging" period. The newness period generally is considered to be the first five to seven years of an aircraft's life until its first D check.[3] The second period, or the mature period, ends at the second D check. It is the most comprehensive scheduled maintenance event in a commercial aircraft's life. The aging period is the period after the second D check to the end of the aircraft's operational life. These three periods combine to make what Boeing terms a "maturity curve."

Maturity curves normalize the costs of aircraft of various ages so that their values can be equitably compared. The mature period is assumed to be the comparison period—Boeing sets the "maturity factor" to be equal to one in the mature period. If aircraft are cheaper

---

[3] A *D check* is a complete structural check and restoration.

to maintain during the newness period, the factor is less than one; if they are more expensive to maintain during the aging period, the factor is greater than one. These maturity factors are used to adjust estimated maintenance costs of different-age aircraft. A rising maturity curve is consistent with age effects, i.e., real maintenance costs rising with age.

Figure 2.1 is adapted from a presentation by Boeing to RAND (Boeing, 2004a). It demonstrates estimates of the maturity factors for two different categories of aircraft types. Boeing aggregated the aircraft to the design type, estimating different maturity curves for two structurally different types of aircraft that it labeled "pre-1980" and "post-1980." In the briefing to RAND, Boeing representatives said that simply assuming one age effect for all aircraft types, as Boeing has done in the past, is no longer viable.

Figure 2.1
Boeing's Maturity Curve

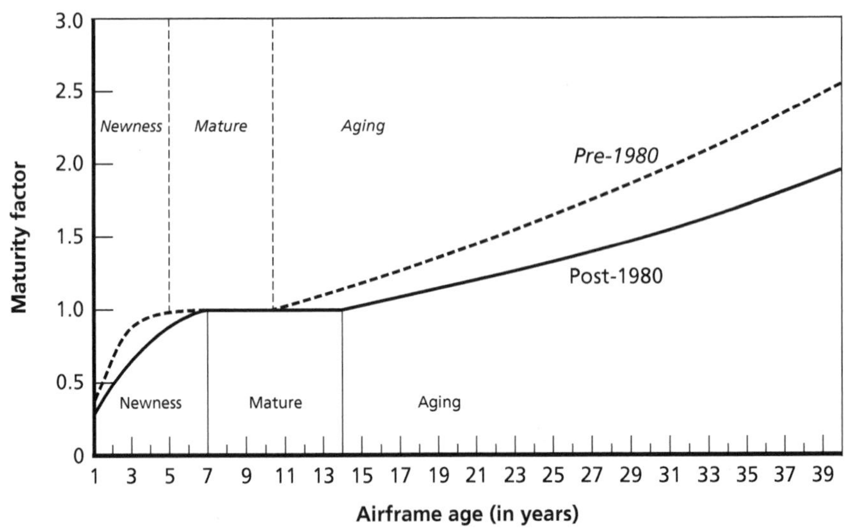

RAND *MG486-2.1*

SOURCE: Modified from Boeing (2004a).

Boeing's previous approach overestimated the cost of future maintenance for new aircraft but underestimated it for older aircraft. The new method will account for differences in aircraft types and eras by first normalizing the current cost of an aircraft and then estimating the future maintenance cost. The accuracy of this method has not yet been proven. Perhaps Boeing's most interesting finding is that, in both eras of aircraft production (pre-1980 and post-1980), the age effect is positive during the aging period, i.e., the curve is convex. Boeing's research provided this study with baseline expectations on the level and shape of age effects.

There has been no formal publication of Boeing's maturity-curve findings, nor was its statistical analysis presented in the briefing. The standard errors, sample sizes, specific model, and assumptions were not supplied.

Figure 2.1 is based on airframe costs. As discussed in Chapter Three, this study uses a more inclusive tabulation of aircraft maintenance costs, including engine costs and overhead along with airframe costs. This study might be viewed as an intellectual sequel to Boeing's work in which a broader swath of aircraft maintenance costs are examined and statistical details not provided by Boeing are supplied.

The next chapter presents a simplified description of military and commercial aircraft maintenance practices and discusses the commercial aviation maintenance data that are central to this study's inquiry.

CHAPTER THREE
Commercial Aviation Maintenance Data

Military and Commercial Aviation Maintenance

Perhaps due to overlap in some of their manufacturers (e.g., Boeing, McDonnell Douglas), commercial and military aircraft are similarly maintained for the most part (although different nomenclature is used to describe various steps in the maintenance process).

In the ordinary course of operation, military aircraft receive installation maintenance, largely from uniformed military maintenance personnel who would then deploy with the aircraft in case of contingency. Some maintenance is done directly on aircraft on the flight line. Other maintenance is done in installation back shops, e.g., a broken part is removed, fixed, and then returned to the aircraft. While still at the home installation, an aircraft might periodically be inspected, repaired, and/or modified by a government-employed civilian depot maintenance team and/or by contractor personnel.

On a calendar basis (e.g., every five or six years), aircraft typically visit an Air Force depot (or contractor facility, varying by aircraft) for PDM or other depot-level maintenance. An aircraft is disassembled during PDM (e.g., the wings and engines are removed), detailed inspection and repair is undertaken, and the aircraft is re-assembled, tested, and returned to its owner. The PDM process is typically lengthy, routinely exceeding 100 days (and sometimes much longer) for many aircraft.

There is also the possibility of unscheduled depot-level maintenance, which is typically caused by a mishap (e.g., two aircraft collide on a runway) or the revelation of a previously unknown flaw causing an entire fleet to be grounded, inspected, and, potentially, repaired. For example, a fatal KC-135E crash in Germany in January 1999 resulted in the remaining KC-135s having flight restrictions and a February 2000 standdown of part of the fleet to search for problems in the stabilizer portion of the tail section (Bolkcom, 2003).

As stated above, commercial aviation maintenance is similar to military aircraft maintenance, but it uses different terminology.[1] At commercial airports, maintenance personnel routinely do a "walk-around" inspection of an aircraft's exterior to look for fuel leaks, worn tires, cracks, dents, and other damage. Every three to five days, there is an *A check* inspection of the aircraft's landing gear, control surfaces, fluid levels, oxygen systems, lighting, and auxiliary power systems. A *B check* occurs every eight months, covering A check topics plus internal control systems, hydraulic systems, and cockpit and cabin emergency equipment. A and B checks can typically be done at airport gates.

A *C check* occurs every 12 to 17 months, during which the aircraft is opened up extensively for inspection for wear, corrosion, and cracks. A C check would typically take place in a maintenance hangar, perhaps at an airline's hub airport. Finally, a *D check* involves the disassembly of an aircraft at a specialized facility. D checks occur on a flying-hour basis (e.g., 747 D checks are to occur after every 22,500 flight hours [Boeing, 1999]), but both their calendar frequency and tasks correspond closely to military PDMs. (D checks, however, tend to be completed more quickly than PDM visits, e.g., in 30 days.)[2] Indeed, both commercial airlines and the Air Force have tried to extend durations between D checks/PDM visits in the interest of maximizing aircraft availability and minimizing lifecycle cost.

Like military aircraft, commercial aircraft sometimes need unscheduled maintenance, but it is less common, because commercial aircraft have lower accident rates and are operated in more-benign con-

---

[1] This discussion is based on Boeing (no date); Aviation Safety Alliance (no date); and Wikipedia (no date).

[2] The *Airline Business* article "Maintenance: Turning the Screw" (2005), for instance, discusses 747-400 D checks being accomplished in four weeks (down from six).

ditions. For example, on May 25, 1979, an American Airlines DC-10 crashed at Chicago's O'Hare airport, killing all on board. Paralleling what happened to the Air Force's KC-135 fleet 20 years later, commercial airlines' DC-10 fleets were grounded while a flaw that caused an engine to fall off the left wing was diagnosed and rectified on the remaining aircraft (Witkin, 1979).

Commercial aircraft have warranties covering their first few years of operation. (The military also makes use of warranties, although Peters and Zycher [2002] found the value of DoD's engine warranties, in particular, to be limited.) In theory, a commercial aircraft owner should not face substantive maintenance costs in its first three to four years of ownership. In practice, as RAND learned from Boeing, aircraft owners do face costs for buying up-front parts inventories, only a portion of which are then reimbursed by manufacturers under warranty. However, what was expected from this current study (and what was found) was that commercial airlines' aircraft maintenance costs jump as the aircraft comes off warranty, and the aircraft's maintenance costs are entirely borne by the aircraft's owner. This phenomenon is not an "aging effect" in the sense of the aircraft degrading markedly. Instead, it represents a new financial burden from the perspective of the airline's owner (relieving a burden previously borne by the aircraft's manufacturer—a burden not observed in the data described and analyzed later in this chapter).

Next, we discuss the DoT data on airline maintenance costs that are key to this study.

Department of Transportation Form 41 Data

Form 41 data are reports that U.S. commercial airlines are required to file with the DoT indicating the airlines' maintenance costs and flying hours. RAND gathered Form 41 data from the DoT on maintenance costs going back to the 1960s. RAND initially gathered the Form 41 longitudinal data in the late 1990s. After collecting and consolidating those RAND data files, the author of this document was able to update the files through the year 2003 and add average fleet ages to each line of data.

The complete panel data used in this analysis date from 1965 to 2003. There are no data for 1985 due to problems in the transition of data-collection methods. Pre-1985 data were collected at the annual level, and post-1985 data were collected at the quarterly level. These data have been combined to form one dataset consisting of annual observations. When aggregated to the airline and aircraft type (737, A310, DC-10, etc.) level, there are 1,003 observations covering nine U.S. airlines (although only American, Delta, Northwest, and United are in our data before 1998) and 29 aircraft types spanning 38 years. Table 3.1 lists the airlines and aircraft types used in this analysis.

Table 3.1
Airlines and Aircraft Types in the Study Data, Earliest Observations, Latest Observations, and Total Observations

	Earliest Observation	Latest Observation	Total Observations
Airline			
Alaska	1998	2003	12
America West	1998	2003	22
American	1965	2003	221
Continental	1998	2003	35
Delta	1965	2003	226
Northwest	1965	2003	203
Southwest[a]	1998	2003	6
United	1965	2003	238
U.S. Airways	1998	2003	40
Aircraft Type			
Boeing 707	1965	1981	30
Boeing 720	1965	1973	24
Boeing 727	1965	2003	142
Boeing 737	1968	2003	94
Boeing 747	1970	2003	88
Boeing 757	1986	2003	84
Boeing 767	1982	2003	70

Table 3.1—Continued

	Earliest Observation	Latest Observation	Total Observations
Aircraft Type (cont.)			
Boeing 777	1995	2003	25
Airbus A300	1988	2003	16
Airbus A310	1991	1995	4
Airbus A319	1998	2003	19
Airbus A320	1989	2003	36
Airbus A321	2001	2003	3
Airbus A330	2000	2003	5
British Aerospace 111	1966	1971	6
Convair CV880	1965	1973	9
Convair CV990	1965	1967	3
Convair CVR580	1986	1988	3
McDonnell Douglas DC-10	1972	2003	95
McDonnell Douglas DC-8	1965	1991	49
McDonnell Douglas DC-9	1966	2003	46
Fokker 100	1991	2003	9
Lockheed L1011	1974	2001	30
Lockheed L188	1965	1970	10
McDonnell Douglas MD-11	1991	2003	24
McDonnell Douglas MD-80	1986	2003	58
McDonnell Douglas MD-90	1995	2003	11
Aerospatiale-Sud Aviation SE210	1965	1970	6
Vickers Viscount 700	1965	1968	4

[a]Southwest Airlines flies only 737s, so RAND had only six observations for Southwest in our data. U.S. Airways, by contrast, flew eight different aircraft types (737s, 757s, 767s, A319s, A320s, A321s, A330s, and MD-80s) during the same six-year period.

The sample does not represent the entirety of U.S. commercial aviation. A number of now-defunct airlines (e.g., Braniff, Eastern, Midway, Pan Am, People Express, Piedmont, TWA), for instance, are not in these data. Gathering the pre-1998 data from DoT paper archives was a very arduous and labor-intensive task, so only the four large airlines (American, Delta, Northwest, and United) were examined for 1965–1997.

Data were aggregated to the type level rather than a separate analysis, for instance, for 737 different variants, because the fleet-age data obtained were at the type level.

An observation consists of three kinds of fleet-level[3] (airline and type of aircraft) variables: categorical variables describing the fleet, continuous variables recording maintenance costs, and continuous variables recording usage. The categorical variables describing a fleet are airline, aircraft type, model, division, and year. The variables recording maintenance costs are separated down to the level of labor or material costs and then to the level of airframe, engine, contracted work, or overhead costs. Finally, the four usage variables are gallons of fuel, flight hours, days assigned to an airline (operations days), and block hours (self-powered hours, including flight time). Table 3.2 provides a sample of three observations.

Noticeably absent from the list of variables are the fleet inventory and average age of the fleet. Average ages were acquired separately from the Securities and Exchange Commission, individual airlines, and Boeing. Fleet inventory was partially available from the airlines and DoT.

The average ages are distributed well from 0 to 25. The oldest recorded average fleet age is 33.5 years (Northwest's Convair 580s in 1988, inherited in their merger with Republic Airlines), but average ages beyond 25 are sparsely represented. Figure 3.1 presents the distribution of average fleet ages.

Because the ages are under-represented beyond 25 years, these results cannot be directly applied to fleets that are much older on average than 25 years.

[3] For purposes of this report, a "fleet" is all individual tail numbers of the same aircraft type flown by an airline, e.g., all American Airlines 747s are a fleet, American Airlines 727s are a separate fleet, and Delta Airlines 727s are a third fleet.

Table 3.2
Sample Observations

	American Airlines	Delta	United
Year	1992	1990	1997
Quarter	1	2	4
Type	727	767	A320
Model	200	200	100/200
Labor costs			
Airframe	$9,613	$746	$2,621
Engine	$3,614	$299	$134
Outside repair costs			
Airframe	$966	$367	$622
Engine	$1,871	$946	$2,298
Materials costs			
Airframe	$8,021	$878	$1,487
Engine	$6,645	$1,471	$65
Total direct airframe costs	$18,600	$1,991	$4,730
Total direct engine costs	$12,130	$2,716	$2,497
Total direct maintenance costs	$30,730	$4,707	$7,227
Maintenance burden costs	$25,379	$2,366	$11,969
Total fleet equipment maintenance costs	$56,109	$7,073	$19,196
Total flight hours	71,151	12,812	36,649
Plane operation days	10,017	1,312	3,657
Block hours	87,707	14,949	41,371
Fuel (gallons)	107,616,515	20,068,729	34,143,558

NOTE: Dollar amounts are expressed in thousands of then-year dollars.

To correct for economy-wide inflation, we used the Bureau of Economic Analysis (2004) implicit price deflators for gross domestic product to translate all maintenance costs into constant year-2000 dollars.

Figure 3.1
Distribution of Average Fleet Ages

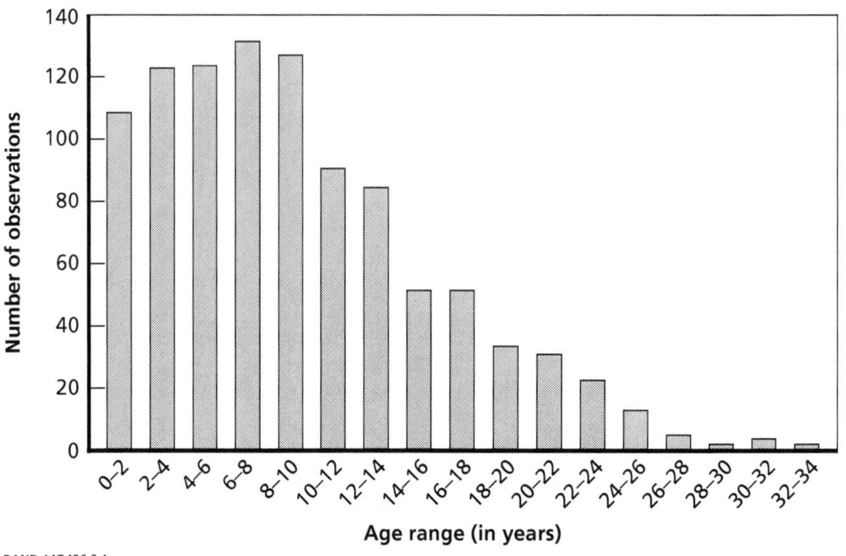

RAND *MG486-3.1*

Estimation Strategy

The goal in this analysis was to estimate how maintenance costs change as aircraft grow older, i.e., "aging effects."

Conceptually, one might think of a regression in which maintenance costs are the dependent variable, and one has a variety of independent variables including fleet age. Maintenance costs are labeled y_{irt}—airline *i*'s annual maintenance costs per flight hour for aircraft type *r* in year *t*. Comparably, Age_{irt} is airline *i*'s average fleet age for its type *r* aircraft in year *t*. The coefficient that one computes for the fleet age variable estimates the aging effect.

A number of statistical issues must be addressed to undertake this estimation.

- The natural logarithm of maintenance costs per flight hour—$\ln(y_{irt})$—is used as the dependent variable. One virtue of the natural log formulation is that the regression coefficient estimates on the age variable are then in percentage terms (which accords with how the existing literature generally presents such results). For instance, a regression coefficient estimate of 0.03 on the age variable (Age_{irt}) suggests that maintenance costs will increase 3 percent for each year of aircraft fleet age, other things being equal.

- One must decide what other independent variables to include in the regression. The Form 41 data allow one to form a number of dichotomous (or "dummy") independent variables. For instance, the type of aircraft involved is known. A 747 is likely to have a different level of maintenance costs per flight hour than the smaller 727, so aircraft-type dummy variables labeled μ_r are included. There also may be industry-wide year effects that might change every aircraft's maintenance cost trends. Technology improves over time, plus the U.S. commercial airline industry was deregulated in 1978. Therefore, calendar-year dummy variables labeled γ_t are included.
- Age break points must be chosen. If one were to simply estimate the regression, $\ln(y_{irt}) = \alpha + \beta \times Age_{irt} + \mu_r + \gamma_t + \varepsilon_{irt}$, one would compute a single age effect β, e.g., maintenance costs, on average, increasing 3 percent per year as aircraft age. Such a single age-effect point estimate would obscure the fact that one expects different age effects (β estimates) in different epochs of an aircraft's life. Boeing, as discussed in Chapter Two, divides aircraft life into "newness," "mature," and "aging" periods and computes three separate aging effects. The same is done here. Form 41 data are divided into three relatively equal-sized categories—fleets with an average age of six years or less, fleets older than six years but no more than 12 years old, and fleets older than 12 years. Then, the regression is estimated separately, $\ln(y_{irt}) = \alpha + \beta \times Age_{irt} + \mu_r + \gamma_t + \varepsilon_{irt}$, thereby computing three different β estimates. These three different β's are different age-effect estimates for different epochs of an aircraft's life.

The output of the estimations is a "maturity curve" estimate. Following Boeing's lead, we used a maturity curve to show how one would expect an aircraft's annual maintenance costs to change as the aircraft ages. Given RAND's econometric approach, the maturity curve will have three segments: ages 0–6 years, ages 6–12 years, and after age 12.

In the next chapter, we present the findings of this study.

CHAPTER FOUR

Study Results

As discussed in Chapter Three, RAND undertook three log linear regressions of the form, $\ln(y_{irt}) = \alpha + \beta \times Age_{irt} + \mu_r + \gamma_t + \varepsilon_{irt}$, where μ_r are aircraft-type dummy variables, γ_t are calendar-year dummy variables, and Age_{irt} are airlines' average fleet ages. Three separate regressions cover fleet ages 0–6 years, 6–12 years, and 12 years and older. The appendix presents the full regression results.

For pedagogical convenience, the three age estimates are presented graphically in this chapter. Figure 4.1 normalizes the total maintenance costs per flight hour for a six-year-old aircraft to 1.0 and depicts how one would expect costs per flight hour to differ relative to that point (which parallels how Boeing set its mature-period "maturity factor" shown in Figure 2.1 to be equal to 1.0).

We estimated a 17.6 percent age effect (with a standard error of 1.8 percent) for the newest aircraft, a 3.5 percent age effect (with a standard error of 0.8 percent) for the mature aircraft (ages 6–12), and a 0.7 percent age effect (with a standard error of 0.7 percent) for the oldest aircraft. In Figure 4.1, the different age curves are artificially linked, e.g., the top of the newness curve connects with the bottom of the mature curve. Boeing used the same approach in constructing its maturity curve (see Figure 2.1). As shown in the appendix, the curves' regressions were estimated separately without imposing this pedagogically useful "curve linkage" constraint. There are, in fact, three separate and independent log linear regressions that created Figure 4.1.

Figure 4.1
Effects of Age on Total Maintenance Costs, Estimated with Form 41 Data

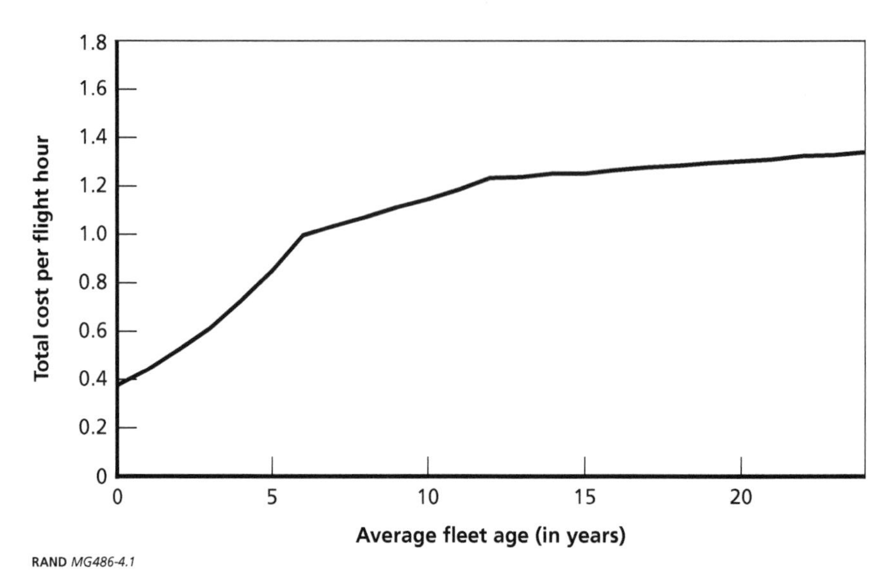

RAND *MG486-4.1*

RAND did not attach great importance to the 17.6 percent aging effect estimate for the newest aircraft. This computed increase in airline maintenance cost is almost certainly a result of aircraft coming off warranty during their third or fourth year of operation. The actual maintenance undertaken on such aircraft presumably does not grow so rapidly. Instead, there are maintenance costs borne early in this period by the aircraft's manufacturer that are not tabulated in the DoT data.

Perhaps the most interesting aspect of Figure 4.1 is the 0.7 percent aging effect estimate for older aircraft. Indeed, one cannot statistically reject a null hypothesis of no maintenance cost growth at all for older aircraft. The concavity of Figure 4.1 is in sharp contrast to the convexity of Figure 2.1. RAND's findings do not support an assumption that aircraft maintenance costs grow rapidly as aircraft age. Such pessimism is not supported by these data, although RAND's airline data are sparse for aircraft past 25 years of age. A more pessimistic maintenance-cost growth pattern may hold for very old aircraft, but U.S. commercial airlines do not hold on to aircraft of such advanced age.

Reconciliation with Boeing's Findings

A key difference between RAND's and Boeing's estimations is that the RAND study team used total maintenance costs per flying hour as the dependent variable, whereas Boeing used airframe maintenance costs per flying hour. As shown in Figure 4.2, airframe maintenance costs in the RAND data were, in total, about a third of total maintenance costs. The rest of the maintenance costs were for the engine and for the "burden," a catchall category including airline overhead, cost of acquiring and maintaining equipment and tools, and other indirect costs.

To more closely parallel Boeing's approach, RAND again ran the three log linear regressions but using only airframe maintenance costs as the log dependent variable. As shown in Figure 4.3, the results are more similar to Boeing's maturity curve in Figure 2.1. In parallel to Figure 4.1, the estimated airframe costs per flight hour of a six-year-old aircraft were normalized to 1, and cost changes are depicted relative to that point. Of special interest, a statistically significant 3.1 percent aging effect was found for aircraft older than 12 years.

Figure 4.2
Average Composition of Total Maintenance Costs

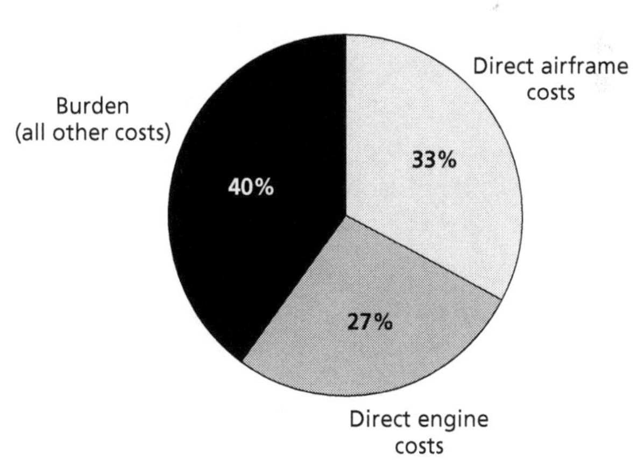

RAND *MG486-4.2*

Figure 4.3
Effects of Age on Airframe Maintenance Costs, Estimated with Form 41 Data

RAND *MG486-4.3*

Figure 4.3 is RAND's version of Boeing's Figure 2.1. Although the results are somewhat different (e.g., the mature phase has 2.3 percent, not 0 percent as in Boeing's version, maintenance cost growth), they are broadly similar, most especially with respect to more rapidly growing airframe maintenance costs for aging aircraft. Therefore, RAND's results broadly accord with Boeing's, but the RAND study team believed that it is more valuable to track total maintenance costs (as in Figure 4.1) than airframe maintenance costs alone (Figure 4.3).

Given the differences between Figures 4.1 and 4.3 (a 0.7 percent versus a 3.1 percent aging effect estimate for the oldest aircraft), it is not surprising that separate regressions for engine and burden costs find very flat regions for older aircraft, as shown in Figures 4.4 and 4.5.

Figure 4.4, in particular, shows sharp growth in engine maintenance costs as engines come off warranty, then considerable flatness thereafter. The concavity of Figures 4.4 and 4.5 offset the convexity of Figure 4.3, making the depiction of overall maintenance costs concave (see Figure 4.1).

Figure 4.4
Effects of Age on Engine Maintenance Costs, Estimated with Form 41 Data

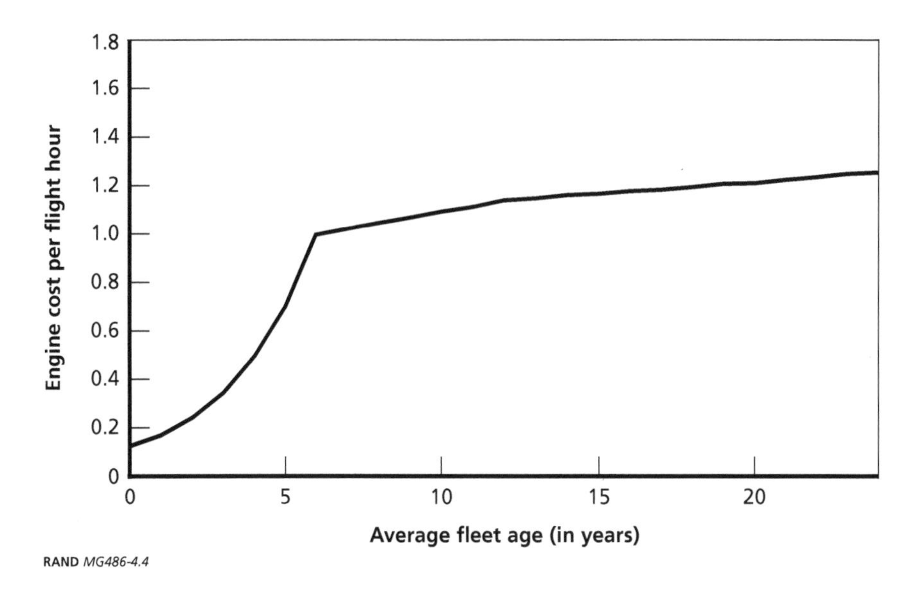

RAND *MG486-4.4*

Figure 4.5
Effects of Age on Maintenance Burden, Estimated with Form 41 Data

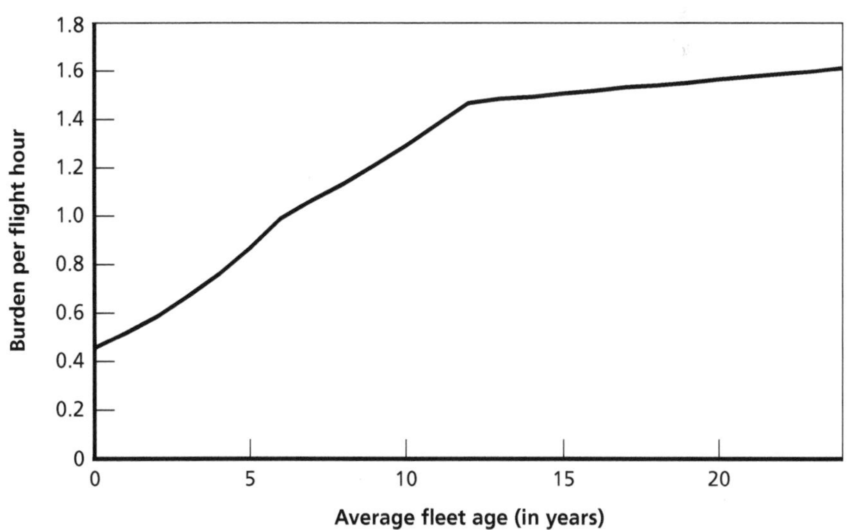

RAND *MG486-4.5*

Alternative Specifications

There are different ways in which the regressions could be set up. In this section, we discuss alternative specifications and their results.

Airline Dummy Variables

It would be straightforward to include airline dummy variables (that would be labeled τ_i) in the regressions. The RAND study team found, however, that in the total maintenance cost regressions, such coefficients were generally not statistically significant. As shown in the appendix, some airline dummy variables were found to be statistically significant in the airframe cost regressions. In general, however, airline dummy variables do not seem to be useful. As shown in Table 4.1, the estimated age effects for total maintenance costs do not meaningfully change when airline dummy variables are included.

Aircraft and Airline Age Interactions

A more troubling possibility would be that different airlines or different types of aircraft have different aging effects. If that were so, it would be harder for the Air Force to generalize from commercial age effects to its situation: There would be no *a priori* way to know whether

Table 4.1
Estimated Age Effects for Total Maintenance Costs per Flight Hour

	Average Age of Airline Fleets		
	0–6 Years	6–12 Years	More than 12 Years
Age effects as shown in Figure 4.1			
Coefficient	17.6%	3.5%	0.7%
Standard error	1.8%	0.8%	0.7%
Age effects with airline dummy variables included			
Coefficient	16.6%	3.3%	0.4%
Standard error	1.6%	0.8%	0.7%

the Air Force's fleets age more like United's or American's or more like a DC-10 or a 727, for example.

To test for such a possibility, the estimations were redone to include age*type ($Age_{irt} \times \mu_r$) and age*airline ($Age_{irt} \times \tau_i$) independent interaction variables. The results, presented in the appendix, showed no systematic pattern. For new aircraft (six years old and younger), five of 25 types had an aging effect statistically significantly greater than the omitted aircraft type (727s). These "poorly aging" young aircraft were the A300, the A319, the A321, the MD-80, and the MD-90. Among the four large airlines for which RAND examined pre-1998 data (see Chapter Three), Delta and United have lower aging coefficients than American has.

For aircraft ages 6–12, the results suggested that the L188 had greater maintenance cost growth than the 727, but the F100 and MD-11 had statistically significantly lower maintenance cost increases in this age range. None of the age*airline interaction terms was statistically significant in this age range.

For aircraft older than 12 years, none of the age*type interaction terms was statistically significant. Meanwhile, United had a greater age coefficient than American, the omitted airline. Table 4.2 summarizes these findings.

Few systematic patterns were found from the use of the interaction terms, as shown in the table. No aircraft type or airline consistently has a statistically significantly greater or lesser age effect across different-aged aircraft. The baseline assumption is that all types of aircraft and airlines face broadly comparable aging effects. At least for the data in this sample, one cannot reject this null hypothesis.

If it is true, this "non-finding" is good news for the Air Force. If, for instance, Delta's aircraft aged significantly differently than American's aircraft, the Air Force would have to identify which airline its maintenance practices more closely resemble to know which airline to more closely examine for "lessons learned." Such a difficult inquiry does not appear to be necessary.

Table 4.2
Results of Search for Type-Specific and/or Airline-Specific Age Effects

	Aircraft Age 0–6 Years	Aircraft Age 6–12 Years	Aircraft Age 12 Years and Older
Age*type interactions	A300, A319, A321, MD-80, and MD-90 age coefficients are significantly greater than the age*727 interaction value	The L188 age coefficient is significantly greater than the 727's; the F100 and MD-11 age coefficients are significantly less than the 727's	No statistically significant age*type coefficients
Age*airline interactions	Delta's and United's age coefficients are significantly less than American's	No statistically significant age*airline coefficients (relative to the omitted American Airlines)	United's age coefficient is significantly greater than American's

Endogenous Selection of Age Breaks

In the baseline estimation, age effects were estimated separately for aircraft ages 0–6 years, 6–12 years, and more than 12 years. A more statistically elegant approach would be to let the data determine the most appropriate or homogeneous splits. Regression tree estimations were used to evaluate different, prospective age break points in the data. The best break points minimize the sum of squared errors.

Regression tree analysis found that the first and most significant split in the data is at age three, when many aircraft warranty periods end. The second most significant split was at 12 years, roughly at the time of the second D check.

However, there is little meaningful difference between the baseline split of 6–12 years and the split of 3–12 years preferred by the regression tree analysis. The 3–12 split yields a slightly higher F statistic, but the 6–12 split yields a slightly larger combined R^2 value. RAND chose to stick with the 6–12 split given its similarity to Boeing's approach. Of course, the age-effect estimate for the post–12-years-old aircraft is the same whether a 3–12-years or 6–12-years split is used.

The study team was heartened by the regression tree results. The results lend credibility to the data and the estimation approach in that the endogenously determined breaks correspond to known events in the life of aircraft. These findings also support the break points Boeing chose to use, shown in Figure 2.1.

CHAPTER FIVE

Potential Bias in Estimated Age Effects

This chapter investigates a potential bias in the commercial aviation data that could diminish the relevance of the commercial findings for the Air Force. In particular, there was concern that commercial airlines might be able to retire "poorly aging" fleets early, whereas the Air Force would not be able to do so. If commercial airlines can dispose of fleets with rapidly increasing maintenance costs (whereas the Air Force cannot), the nearly flat region depicted in Figure 4.1 after the 12-year mark in average fleet age might not be representative of what the Air Force should anticipate.

Indeed, some commercial fleets have been retired earlier than what is normal, where retirement at an average age greater than 20 years is defined as "normal." Over the past 40 years, 21 of 82 fleets in the DoT Form 41 data used in this study were retired (or, more likely, sold to foreign or cargo airlines) before reaching an average age of 20 years. RAND labeled these "short-lived fleets." Table 5.1 summarizes those fleets.

The 21 fleets in the table account for 156 of the 1,003 data points used in the analysis in Chapter Four.

To assess whether these short-lived fleets had aging characteristics that differed from those of normally aging fleets, RAND created a dummy variable labeled short_fleets that has a value of 1 if the fleet is one of those listed in Table 5.1 and a value of 0 if it is not. RAND then estimated the model $\ln(y_{irt}) = \alpha_1 + \alpha_2 \times \text{short\_fleets} + \beta_1 \times \text{Age}_{irt} + \beta_2 \times \text{Age}_{irt} \times \text{short\_fleets} + \mu_r + \gamma_t + \varepsilon_{irt}$. This specification allows both the intercept and the age slope to be different for the early-retiring fleets (short_fleets = 1).

Table 5.1
Characteristics of Short-Lived Fleets

Airline	Aircraft Type	Age Period at Retirement	Average Age at Retirement (years)	Year of Retirement
American	707	Aging	14.1	1981
American	720	Mature	11	1971
American	747	Aging	12.7	1992
American	BAC111	Newness	5.6	1971
American	CV990	Newness	5.4	1967
American	L188	Mature	9.1	1968
American	MD-11	Mature	8.8	2001
American	MD-90	Newness	4	2000
Delta	747	Newness	5.5	1976
Delta	A310	Newness	4	1994
Delta	CV880	Mature	12.3	1973
Delta	DC-10	Aging	13.9	1988
Northwest	707	Mature	9.4	1977
Northwest	720	Mature	9.5	1973
Northwest	L188	Mature	11	1970
Northwest	MD-80	Aging	17.5	1999
United	720	Mature	11.7	1972
United	L1011	Mature	7.4	1988
United	SE210	Mature	9.2	1970
United	V700	Aging	12.1	1968
U.S. Airways	MD-80	Aging	19	2001

As shown in Table 5.2, none of the short_fleets intercepts or age coefficients is statistically significantly different from zero. RAND found no evidence that the cost per flight hour of these short-lived fleets increased any differently than that of other fleets that lived a full life or are still in service today.

Table 5.2
Estimation Model with Short-Lived Fleet Intercepts and Aging Effects

	Average Age of Fleets		
	0–6 Years	6–12 Years	More than 12 Years
α_2, short-lived fleets intercept (difference between the intercepts for non-early-retired aircraft and for short-lived fleets)	0.12521	0.14985	0.17352
α_2 standard error	0.17579	0.19865	0.45986
T-statistic	0.71	0.75	0.38
Probability	0.4769	0.4513	0.7057
β_2, short-lived fleets aging effect (difference between the aging effects observed for non-early-retired aircraft and for short-lived fleets)	0.03463	0.00091	–0.00937
β_2 standard error	0.04359	0.02139	0.03022
T-statistic	0.79	0.04	–0.31
Probability	0.4276	0.9661	0.7567

So why were these short-lived fleets retired early? Table 5.3 suggests one hypothesis. Instead of running a model with different intercepts and age effects for short-lived fleets, RAND ran the following:

$$\ln(y_{irt}) = \alpha_1 + \alpha_2 \times \text{short\_fleets} + \beta \times Age_{irt} + \mu_r + \gamma_t + \varepsilon_{irt},$$

where only the intercept varies and short-lived fleets are assumed to have typical aging effects.

With this specification, the short_fleets intercept term is positive and significant during both the newness and mature periods (see Chapters Two and Three for a discussion of the different periods of aircraft life). Further, RAND had only 20 short-lived-fleet observations for fleets in the 12-plus-years-of-life aging period. One might infer that the short-lived fleets were unusually expensive and that their elevated

Table 5.3
Different Intercept Only, for Short-Lived Fleets

	Average Age of Fleets		
	0–6 Years	6–12 Years	More than 12 Years
α_2, short fleets intercept (difference between the intercepts for non-early-retired aircraft and for short-lived fleets)	0.23489	0.15799	0.03360
α_2 standard error	0.10876	0.05424	0.08407
T-statistic	2.16	2.91	0.40
Probability	0.0316	0.0039	0.6897

maintenance costs motivated the airlines to replace them. But their maintenance costs did not grow abnormally—they started out abnormally high.

Of course, there may be other possible reasons for these short-life-span fleets not surviving to 20 years of life. Those reasons may include loss of consumer demand for a particular type of aircraft, regulatory changes, an airline's shifting business plan, and increased operating costs. Unusual maintenance-cost aging effects do not appear to be a contributor to the short life spans.

RAND found no evidence that airlines retired certain fleets early because the aircraft were aging abnormally poorly. Airline fleet retirement does not appear to be causing Figure 4.1's concavity.

As noted in Chapter Three, this analysis covers only four major airlines (American, Delta, Northwest, and United) for 1965–1997. It could be that now-vanished airlines had greater problems with aging aircraft (which perhaps even contributed to the airlines' demise). Testing this hypothesis would require gathering and tabulating more of the older DoT Form 41 data.

CHAPTER SIX

Conclusions

Readers initially may have been curious about why a study of fleets belonging to commercial airlines would be relevant to the Air Force's military aircraft. This monograph presents research funded by the U.S. Air Force and undertaken by an Air Force officer while in the Pardee RAND Graduate School. Yet, the subject of this study is how the costs of maintaining commercial aircraft change as those aircraft age.

For this research to be relevant to the Air Force, one must accept the argument that the commercial airlines' experience is meaningfully analogous to the Air Force's experience. The analogy between commercial aviation and military aviation is closest for the Air Force's executive transport aircraft, which are, for the most part, COTS. But more importantly, Air Force tankers and cargo aircraft are similar, at least across some dimensions, to commercial passenger aircraft. But the analogy is open to questioning. No commercial airliners in the data used in this study carry fuel for air-to-air refueling of other aircraft, for instance. In addition, commercial airliners fly many more hours per year than any military aircraft, even during combat periods. One might wonder how maintenance needs would evolve for a commercial aircraft that flew only 500 hours per year (as opposed to the thousands of hours per year that commercial aircraft commonly fly), but no profitable airline would operate an aircraft on such a limited basis.

As stated earlier, this study used DoT data from 1965–2003 to analyze how total commercial aircraft maintenance costs (correcting for economy-wide inflation) appear to change as commercial aircraft age. In accord with related research by Boeing (Boeing, 2004a), RAND found that airline maintenance costs in the first six years of aircraft ownership grow at a fairly sharp rate. This increase in maintenance costs early in an aircraft's life is, essentially, cost-shifting (which is unobserved in the data) from manufacturer-provided maintenance under aircraft warranties to maintenance paid for by the airlines.

For commercial fleets averaging 6 to 12 years of age, RAND found that maintenance costs per flying hour increase about 3.5 percent per year of aircraft age, other factors being equal.

Perhaps the most intriguing result of this analysis is that total maintenance costs grow only slightly (not statistically significantly different from zero) for aircraft more than 12 years old. As Boeing had found earlier, RAND found that airframe maintenance costs grow for these older aircraft, but when engine and overhead costs are also factored in, there is very little growth in total maintenance costs per flying hour for those aircraft. Engine maintenance costs, in particular, remain very flat as aircraft age (after an initial jump in maintenance costs in the first few years of ownership).

Even if one accepts that the commercial aviation experience is relevant to the Air Force experience, there are limits as to what to infer from this analogy. U.S. airlines generally do not operate aircraft that are more than 25 years old. Hence, the flatness of maintenance costs late in the life of commercial aircraft says little about what might happen to, for instance, maintenance costs for military aircraft such as the B-52 or KC-135, given that those aircraft passed 25 years of service many years ago. (By the same token, these very old military aircraft may have relatively few lifetime flight hours by commercial aviation standards.)

This study suggests that pessimism about the future trajectory of total maintenance costs for military aircraft systems is not necessarily warranted. The assumption that total maintenance costs always grow rapidly as aircraft age may not be correct. When, for instance, the Air Force models the future maintenance costs of a new system, it may be appropriate for the Air Force to consider the possibility of a midlife (by military standards) period of relative stasis in maintenance costs, at least through the roughly 25-year point in the life of the system. This study also suggests that different types of aircraft maintenance costs, e.g., airframe maintenance versus engine maintenance, may show different cost patterns.

APPENDIX

Regression Results

This appendix provides the statistical output underlying the results presented in Chapters Four and Five. Figure 4.1's concave aging curve, the central result of this analysis, comes from three separate log-linear regressions. Each regression is of the form, $\ln(y_{irt}) = \alpha + \beta * Age_{irt} + \mu_r + \gamma_t + \varepsilon_{irt}$, where μ_r are aircraft-type dummy variables, γ_t are calendar-year dummy variables, and Age_{irt} are the average ages of the airlines' fleets. The three different regressions emanate from dividing the data among fleets six years of age and younger ($Age \leq 6$), fleets ages 6–12 years ($6 < Age \leq 12$), and fleets more than 12 years old ($12 < Age$).

In the tabulation that follows, the intercept value (α) is displayed first. Next, the Age (Age_{irt}) variable of central interest is displayed, followed by aircraft type dummy variables (μ_r, 707-V700), where 727 is the omitted aircraft-type variable. Next are year dummy variables (γ_t), with 1986 as the omitted year (and 1985 is always dropped, because data for 1985 are missing from the DoT data used for this study). A variable was dropped from the estimation if there were no observations of the variable in the subsample. For example, all 777s in the DoT data were quite new, so the 777 dummy variable was dropped from the 6 < Age < = 12 and 12 < Age regressions.

Baseline Regressions

Age ≤ 6

```
Number of obs =      355
F( 63,    290) =   12.19
Prob > F      =   0.0000
R-squared     =   0.7258
Adj R-squared =   0.6662
Root MSE      =   .41296
```

Source	SS	df	MS
Model	130.91233	63	2.07797
Residual	49.45426	290	0.17053
Total	180.36659	353	

	Coef.	Std. Err.	t	P>|t|
Intercept	-1.99624	0.24636	-8.10	0.0000
Age	0.17634	0.01754	10.05	0.0000
707	-0.19895	0.14455	-1.38	0.1698
720	-0.44943	0.20457	-2.20	0.0288
727	(omitted)			
737	0.12438	0.17541	0.71	0.4788
747	1.24815	0.13453	9.28	0.0000
757	0.74363	0.21673	3.43	0.0007
767	0.92877	0.21040	4.41	0.0000
777	1.28132	0.23343	5.49	0.0000
A300	1.24816	0.26243	4.76	0.0000
A310	0.84997	0.29659	2.87	0.0045
A319	0.01997	0.24231	0.08	0.9344
A320	0.56165	0.22479	2.50	0.0130
A321	-0.50183	0.32997	-1.52	0.1294
A330	1.00955	0.31189	3.24	0.0013
BAC111	0.17399	0.18863	0.92	0.3571
CV880	-0.10166	0.32395	-0.31	0.7539
CV990	0.38085	0.26788	1.42	0.1562
CVR580	(dropped)			
DC10	0.95905	0.14835	6.46	0.0000
DC8	-0.26918	0.16887	-1.59	0.1120
DC9	-0.21152	0.16587	-1.28	0.2032
F100	0.85462	0.26955	3.17	0.0017
L1011	1.29734	0.17969	7.22	0.0000
L188	(dropped)			
MD11	1.40350	0.23682	5.93	0.0000
MD80	0.13193	0.23734	0.56	0.5787
MD90	0.74269	0.26126	2.84	0.0048
SE210	-0.38249	0.32265	-1.19	0.2368

V700	(dropped)			
y1965	1.54560	0.27572	5.61	0.0000
y1966	1.35162	0.26732	5.06	0.0000
y1967	1.28541	0.26919	4.78	0.0000
y1968	1.21992	0.26043	4.68	0.0000
y1969	1.11060	0.25892	4.29	0.0000
y1970	0.98323	0.25075	3.92	0.0001
y1971	0.76222	0.25147	3.03	0.0027
y1972	0.76482	0.25027	3.06	0.0025
y1973	0.66215	0.25258	2.62	0.0092
y1974	0.80025	0.25784	3.10	0.0021
y1975	0.65394	0.26421	2.48	0.0139
y1976	0.54703	0.26458	2.07	0.0396
y1977	0.49505	0.28810	1.72	0.0868
y1978	0.31963	0.30023	1.06	0.2879
y1979	0.34767	0.36063	0.96	0.3358
y1980	0.25432	0.36082	0.70	0.4815
y1981	0.05572	0.47266	0.12	0.9062
y1982	0.19651	0.33599	0.58	0.5591
y1983	0.00129	0.28995	0.00	0.9965
y1984	0.00471	0.23909	0.02	0.9843
y1985	(dropped)			
y1986	(omitted)			
y1987	0.04893	0.20870	0.23	0.8148
y1988	-0.04549	0.20910	-0.22	0.8279
y1989	-0.11243	0.19994	-0.56	0.5744
y1990	-0.05448	0.20607	-0.26	0.7917
y1991	0.10936	0.19389	0.56	0.5732
y1992	-0.08349	0.19322	-0.43	0.6660
y1993	-0.09963	0.19123	-0.52	0.6028
y1994	0.01076	0.20903	0.05	0.9590
y1995	-0.14824	0.20949	-0.71	0.4798
y1996	-0.23871	0.21721	-1.10	0.2727
y1997	-0.16446	0.22453	-0.73	0.4645
y1998	-0.45683	0.21904	-2.09	0.0379
y1999	-0.03919	0.19844	-0.20	0.8436
y2000	0.06895	0.20233	0.34	0.7335
y2001	-0.06513	0.20060	-0.32	0.7456
y2002	0.03727	0.20252	0.18	0.8541
y2003	0.07223	0.20856	0.35	0.7293

6 < Age ≤ 12

```
Number of obs =      349
F( 58,    290) =   23.84
Prob > F      =   0.0000
R-squared     =   0.8266
Adj R-squared =   0.7920
Root MSE      =   .19615
```

Source	SS	df	MS
Model	53.20140	58	0.917227
Residual	11.15780	290	0.03848

Total	64.35920	348

	Coef.	Std. Err.	t	P>|t|
Intercept	-1.14539	0.13481	-8.50	0.0000
Age	0.03525	0.00803	4.39	0.0000
707	0.25832	0.06846	3.77	0.0002
720	0.08057	0.10844	0.74	0.4581
727	(omitted)			
737	0.19083	0.06320	3.02	0.0028
747	1.23502	0.05052	24.44	0.0000
757	0.51476	0.07133	7.22	0.0000
767	0.61787	0.07251	8.52	0.0000
777	(dropped)			
A300	1.22277	0.10484	11.66	0.0000
A310	(dropped)			
A319	(dropped)			
A320	0.32978	0.09205	3.58	0.0004
A321	(dropped)			
A330	(dropped)			
BAC111	(dropped)			
CV880	0.45293	0.12873	3.52	0.0005
CV990	(dropped)			
CVR580	(dropped)			
DC10	0.97107	0.05719	16.98	0.0000
DC8	0.22933	0.07829	2.93	0.0037
DC9	0.15920	0.08243	-1.93	0.0544
F100	0.00907	0.15583	0.06	0.9536
L1011	0.87773	0.08784	9.99	0.0000
L188	0.13121	0.14289	0.92	0.3593
MD11	0.92243	0.09988	9.24	0.0000
MD80	0.25483	0.07416	3.44	0.0007
MD90	0.57751	0.13412	4.31	0.0000
SE210	0.11962	0.14880	0.80	0.4221
V700	-0.36247	0.19405	-1.87	0.0628
y1965	0.53170	0.22235	2.39	0.0174
y1966	0.55937	0.21923	2.55	0.0112
y1967	0.68953	0.18074	3.82	0.0002
y1968	0.58849	0.17508	3.36	0.0009
y1969	0.53338	0.17376	3.07	0.0023
y1970	0.58909	0.15915	3.70	0.0003
y1971	0.42169	0.15283	2.76	0.0062
y1972	0.29858	0.14216	2.10	0.0366
y1973	0.41477	0.13404	3.09	0.0022
y1974	0.43949	0.12785	3.44	0.0007
y1975	0.38682	0.12656	3.06	0.0024
y1976	0.36863	0.12561	2.93	0.0036
y1977	0.34858	0.12099	2.88	0.0043
y1978	0.26643	0.11965	2.23	0.0267
y1979	0.16546	0.17750	1.41	0.1602
y1980	0.21641	0.11886	1.82	0.0697
y1981	0.18824	0.11854	1.59	0.1134
y1982	0.00277	0.12002	0.02	0.9816
y1983	0.05272	0.11890	0.44	0.6578
y1984	-0.02115	0.12885	-0.16	0.8697

	Coef.	Std. Err.	t	P>|t|
y1985	(dropped)			
y1986	(omitted)			
y1987	0.09588	0.12703	0.75	0.4510
y1988	0.06011	0.12480	0.48	0.6304
y1989	0.13466	0.12761	1.06	0.2922
y1990	0.17370	0.12537	1.39	0.1670
y1991	0.19320	0.13946	1.39	0.1670
y1992	-0.04096	0.14896	-0.27	0.7835
y1993	-0.04389	0.14858	-0.30	0.7679
y1994	-0.16227	0.13684	-1.19	0.2367
y1995	-0.15570	0.13395	-1.16	0.2460
y1996	-0.16573	0.13148	-1.26	0.2085
y1997	-0.11328	0.13177	-0.86	0.3907
y1998	-0.09962	0.11990	-0.83	0.4068
y1999	-0.06177	0.12032	-0.51	0.6081
y2000	0.06143	0.11980	0.51	0.6085
y2001	0.21006	0.11990	1.75	0.0808
y2002	0.18642	0.12077	1.54	0.1238
y2003	0.04379	0.12258	0.36	0.7212

12 < Age

```
Number of obs =      299
F( 40,   258) =    18.86
Prob > F      =   0.0000
R-squared     =   0.7452
Adj R-squared =   0.7057
Root MSE      =   .24612
```

Source	SS	df	MS
Model	45.70615	40	1.14265
Residual	15.62819	258	0.06057
Total	61.33434	298	

	Coef.	Std. Err.	t	P>|t|
Intercept	-0.50908	0.12704	-4.01	0.0000
Age	0.00715	0.00715	1.00	0.3182
707	-0.02010	0.17553	-0.11	0.9089
720	(dropped)			
727	(omitted)			
737	0.03681	0.06703	0.55	0.5833
747	0.82491	0.06056	13.62	0.0000
757	0.35910	0.11829	3.04	0.0026
767	0.19868	0.13197	1.51	0.1334
777	(dropped)			
A300	0.77283	0.16525	4.68	0.0000
A310	(dropped)			
A319	(dropped)			
A320	(dropped)			
A321	(dropped)			

```
    A330 |  (dropped)
  BAC111 |  (dropped)
   CV880 |  0.26891     0.25834         1.04    0.2989
   CV990 |  (dropped)
  CVR580 | -0.09270     0.19490        -0.48    0.6347
    DC10 |  0.79648     0.04596        17.33    0.0000
     DC8 |  0.27332     0.07655         3.57    0.0004
     DC9 | -0.24742     0.06489        -3.81    0.0002
    F100 |  (dropped)
   L1011 |  0.49307     0.08199         6.01    0.0000
    L188 |  (dropped)
    MD11 |  (dropped)
    MD80 | -0.00932     0.07763        -0.12    0.9045
    MD90 |  (dropped)
   SE210 |  (dropped)
    V700 |  (dropped)
   y1965 |  (dropped)
   y1966 |  (dropped)
   y1967 |  (dropped)
   y1968 | -0.28767     0.25840        -1.11    0.2666
   y1969 |  (dropped)
   y1970 |  (dropped)
   y1971 |  (dropped)
   y1972 |  (dropped)
   y1973 |  (dropped)
   y1974 |  (dropped)
   y1975 |  (dropped)
   y1976 |  (dropped)
   y1977 |  0.20204     0.19932         1.01    0.3117
   y1978 |  0.13867     0.19912         0.70    0.4868
   y1979 |  0.24616     0.17745         1.39    0.1666
   y1980 |  0.23320     0.15339         1.52    0.1297
   y1981 |  0.15542     0.13927         1.12    0.2655
   y1982 |  0.05676     0.13313         0.43    0.6702
   y1983 | -0.07144     0.12522        -0.57    0.5688
   y1984 | -0.13111     0.11371        -1.15    0.2500
   y1985 |  (dropped)
   y1986 |  (omitted)
   y1987 | -0.01428     0.09710        -0.15    0.8832
   y1988 | -0.01216     0.09914        -0.12    0.9024
   y1989 |  0.11553     0.10442         1.11    0.2696
   y1990 |  0.15417     0.10516         1.47    0.1439
   y1991 |  0.20028     0.10052         1.99    0.0474
   y1992 |  0.15722     0.10290         1.53    0.1278
   y1993 |  0.06035     0.10756         0.56    0.5753
   y1994 |  0.04866     0.10462         0.47    0.6422
   y1995 |  0.06010     0.10631         0.57    0.5723
   y1996 |  0.10376     0.10839         0.96    0.3393
   y1997 |  0.11280     0.10777         1.05    0.2962
   y1998 |  0.27348     0.10561         2.59    0.0102
   y1999 |  0.23838     0.10781         2.21    0.0279
   y2000 |  0.26775     0.11088         2.41    0.0164
   y2001 |  0.29075     0.11228         2.59    0.0102
   y2002 |  0.18833     0.11653         1.62    0.1073
   y2003 |  0.23762     0.11717         2.03    0.0436
------------------------------------------------------
```

The results of Boeing's seminal maturity-model, shown in Figure 2.1, were computed by analyzing only airframe maintenance costs, not total aircraft maintenance costs. To check RAND's findings against Boeing's, the RAND study team ran separate regressions for airframe costs, engine costs, and burden costs (all other costs, including overhead). In the airframe case, the study team found it helpful (in the sense of improving goodness-of-fit) to include airline dummy variables (Alaska Airlines through U.S. Airways). American Airlines was the omitted airline in constructing these dichotomous independent airline variables.

Airframe Cost Regressions

Age ≤ 6

```
Number of obs =      355
F( 32,    321) =   23.59
Prob > F      =   0.0000
R-squared     =   0.7016
Adj R-squared =   0.6719
Root MSE      =   .34816

      Source |        SS         df          MS
-------------+------------------------------
       Model |    91.48985     32      2.85906
    Residual |    38.91006    321      0.12122
-------------+------------------------------
       Total |   130.39991    353

------------------------------------------------------------
             |      Coef.     Std. Err.          t     P>|t|
-------------+----------------------------------------------
   Intercept | -1.87536     0.08228       -22.79    0.0000
         Age |  0.07163     0.01210         5.92    0.0000
         707 |  0.22478     0.11730         1.92    0.0562
         720 |  0.22703     0.14818         1.53    0.1265
         727 |  (omitted)
         737 | -0.54595     0.10586        -5.16    0.0000
         747 |  0.90606     0.09710         9.33    0.0000
         757 | -0.31000     0.08533        -3.63    0.0003
         767 |  0.11816     0.08948         1.32    0.1876
         777 |  0.24788     0.09793         2.53    0.0118
        A300 |  0.65845     0.15046         4.38    0.0000
        A310 |  0.63369     0.18964         3.34    0.0009
        A319 | -0.80272     0.12015        -6.68    0.0000
        A320 | -0.26374     0.09994        -2.64    0.0087
        A321 | -1.20979     0.24412        -4.96    0.0000
```

A330	0.62328	0.20846	2.99	0.0030
BAC111	0.34264	0.15965	2.15	0.0326
CV880	0.64639	0.25778	2.51	0.0127
CV990	0.71381	0.21434	3.33	0.0010
CVR580	(dropped)			
DC10	0.58595	0.09838	5.96	0.0000
DC8	0.16110	0.12940	1.24	0.2141
DC9	-0.05545	0.14354	-0.39	0.6996
F100	0.10131	0.15072	0.67	0.5019
L1011	0.72688	0.12683	5.73	0.0000
L188	(dropped)			
MD11	0.45468	0.10722	4.24	0.0000
MD80	-0.52539	0.12060	-4.36	0.0000
MD90	-0.27587	0.14066	-1.96	0.0507
SE210	-0.03726	0.25734	-0.14	0.8850
V700	(dropped)			
Alaska	(dropped)			
America West	0.54692	0.17104	3.20	0.0015
American	(omitted)			
Continental	0.18028	0.10512	1.72	0.0873
Delta	-0.10984	0.06274	-1.75	0.0810
Northwest	-0.23605	0.07011	-3.37	0.0009
Southwest	(dropped)			
United	0.15949	0.06235	2.56	0.0110
US Airways	-0.11766	0.13368	-0.88	0.3794

6 < Age ≤ 12

```
Number of obs =      349
F( 29,    319) =   28.11
Prob > F      =   0.0000
R-squared     =   0.7187
Adj R-squared =   0.6932
Root MSE      =   .25703
```

Source	SS	df	MS
Model	53.85416	29	1.85704
Residual	21.07428	319	0.06606
Total	74.92844	348	

	Coef.	Std. Err.	t	P>|t|
Intercept	-1.87995	0.09164	-20.51	0.0000
Age	0.02257	0.00880	2.56	0.0108
707	0.17852	0.08107	2.20	0.0284
720	0.42314	0.07424	5.70	0.0000
727	(omitted)			
737	0.27273	0.06356	4.29	0.0000
747	1.13689	0.05933	19.16	0.0000
757	0.14365	0.05854	2.45	0.0147
767	0.49539	0.06162	8.04	0.0000

	Coef.	Std. Err.	t	P>|t|
777	(dropped)			
A300	1.05599	0.11551	9.14	0.0000
A310	(dropped)			
A319	(dropped)			
A320	0.22572	0.11018	2.05	0.0413
A321	(dropped)			
A330	(dropped)			
BAC111	(dropped)			
CV880	0.88500	0.11765	7.52	0.0000
CV990	(dropped)			
CVR580	(dropped)			
DC10	0.73812	0.06897	10.70	0.0000
DC8	0.47753	0.08207	5.82	0.0000
DC9	0.28235	0.11074	2.55	0.0112
F100	-0.09483	0.18865	-0.50	0.6155
L1011	1.10988	0.10269	10.81	0.0000
L188	0.67910	0.09219	7.37	0.0000
MD11	0.92926	0.10706	8.68	0.0000
MD80	0.15581	0.07020	2.22	0.0271
MD90	0.83063	0.15824	5.25	0.0000
SE210	0.48311	0.13745	3.51	0.0005
V700	0.19464	0.15630	1.25	0.2139
Alaska	0.41033	0.09234	4.44	0.0000
America West	0.45555	0.14549	3.13	0.0019
American	(omitted)			
Continental	-0.05756	0.11928	-0.48	0.6298
Delta	-0.32579	0.05041	-6.46	0.0000
Northwest	-0.24937	0.04484	-5.56	0.0000
Southwest	0.23486	0.12082	1.94	0.0528
United	-0.08734	0.04554	-1.92	0.0560
US Airways	0.04794	0.10197	0.47	0.6386

12 < Age

```
Number of obs =     299
F( 22,   276) =   19.95
Prob > F      =  0.0000
R-squared     =  0.6139
Adj R-squared =  0.5831
Root MSE      =  .30582
```

Source	SS	df	MS
Model	41.04552	22	1.86571
Residual	25.81295	276	0.09353
Total	66.85847	298	

	Coef.	Std. Err.	t	P>|t|
Intercept	-1.80564	0.12214	-14.78	0.0000
Age	0.03129	0.00581	5.39	0.0000
707	-0.14894	0.18736	-0.79	0.4273

```
         720 |  (dropped)
         727 |  (omitted)
         737 |  0.00870     0.08134          0.11    0.9149
         747 |  0.62362     0.07366          8.47    0.0000
         757 |  0.18106     0.15679          1.15    0.2492
         767 |  0.30366     0.14460          2.10    0.0366
         777 |  (dropped)
        A300 |  0.90570     0.18760          4.83    0.0000
        A310 |  (dropped)
        A319 |  (dropped)
        A320 |  (dropped)
        A321 |  (dropped)
        A330 |  (dropped)
      BAC111 |  (dropped)
       CV880 |  0.19516     0.31263          0.62    0.5330
       CV990 |  (dropped)
      CVR580 | -0.24405     0.19667         -1.24    0.2157
        DC10 |  0.65285     0.05718         11.42    0.0000
         DC8 |  0.16234     0.07604          2.14    0.0336
         DC9 | -0.19167     0.07285         -2.63    0.0090
        F100 |  (dropped)
       L1011 |  0.65236     0.10953          5.96    0.0000
        L188 |  (dropped)
        MD11 |  (dropped)
        MD80 | -0.06376     0.08595         -0.74    0.4588
        MD90 |  (dropped)
       SE210 |  (dropped)
        V700 | -0.30959     0.31218         -0.99    0.3222
      Alaska |  0.13228     0.31671          0.42    0.6765
America West |  0.61840     0.13351          4.63    0.0000
    American |  (omitted)
 Continental |  0.42503     0.10067          4.22    0.0000
       Delta | -0.11003     0.07301         -1.51    0.1329
   Northwest | -0.20826     0.06088         -3.42    0.0007
   Southwest |  (dropped)
      United |  0.00836     0.06082          0.14    0.8908
  US Airways |  0.06446     0.10532          0.61    0.5410
----------------------------------------------------------------
```

Engine Cost Regressions

Age ≤ 6

```
Number of obs =      355
F( 63,    290) =   10.26
Prob > F      =   0.0000
R-squared     =   0.6902
Adj R-squared =   0.6229
Root MSE      =   .73633

Source |          SS         df          MS
```

	SS	df	MS
Model	350.32193	63	5.56067
Residual	157.23201	290	0.54218
Total	507.55394	353	

	Coef.	Std. Err.	t	P>|t|
Intercept	-4.42263	0.43928	-10.07	0.0000
Age	0.42732	0.03127	13.66	0.0000
707	-0.63203	0.25775	-2.45	0.0148
720	-1.56375	0.36477	-4.29	0.0000
727	(omitted)			
737	0.15571	0.31276	0.50	0.6190
747	2.08638	0.23988	8.70	0.0000
757	1.19678	0.38644	3.10	0.0021
767	0.96622	0.37516	2.58	0.0105
777	1.88739	0.41622	4.53	0.0000
A300	1.71066	0.46793	3.66	0.0003
A310	1.07721	0.52884	2.04	0.0426
A319	0.40392	0.43205	0.93	0.3506
A320	1.04236	0.40082	2.60	0.0098
A321	0.63208	0.58837	1.07	0.2836
A330	1.28800	0.55612	2.32	0.0213
BAC111	-0.35531	0.33635	-1.06	0.2917
CV880	-1.18651	0.57762	-2.05	0.0409
CV990	-0.58805	0.47765	-1.23	0.2193
CVR580	(dropped)			
DC10	1.80381	0.26453	6.82	0.0000
DC8	-1.29068	0.30110	-4.29	0.0000
DC9	-0.37871	0.29575	-1.28	0.2014
F100	1.07694	0.48063	2.24	0.0258
L1011	1.77231	0.32040	5.53	0.0000
L188	(dropped)			
MD11	2.08702	0.42226	4.94	0.0000
MD80	0.18605	0.42320	0.44	0.6605
MD90	1.82381	0.46584	3.92	0.0001
SE210	-1.52157	0.57530	-2.64	0.0086
V700	(dropped)			
y1965	2.78558	0.49163	5.67	0.0000
y1966	2.37739	0.47664	4.99	0.0000
y1967	2.17352	0.47999	4.53	0.0000
y1968	1.83856	0.46437	3.96	0.0000
y1969	1.67526	0.46168	3.63	0.0003
y1970	1.20169	0.44710	2.69	0.0076
y1971	0.77659	0.44839	1.73	0.0843
y1972	0.62377	0.44625	1.40	0.1632
y1973	0.47515	0.45036	1.06	0.2923
y1974	0.59933	0.45975	1.30	0.1934
y1975	0.58126	0.47111	1.23	0.2183
y1976	0.04506	0.47177	0.10	0.9240
y1977	0.10557	0.51370	0.21	0.8373
y1978	-0.20853	0.53533	-0.39	0.6972
y1979	-0.14369	0.64303	-0.22	0.8233
y1980	-0.17624	0.64337	-0.27	0.7843

	Coef.	Std. Err.	t	P>|t|
y1981	-0.53713	0.84279	-0.64	0.5244
y1982	-0.63676	0.59910	-1.06	0.2887
y1983	-0.84798	0.51701	-1.64	0.1021
y1984	-0.21568	0.42631	-0.51	0.6133
y1985	(dropped)			
y1986	(omitted)			
y1987	0.22866	0.37213	0.61	0.5394
y1988	0.33019	0.37284	0.89	0.3766
y1989	-0.26958	0.35651	-0.76	0.4502
y1990	-0.29095	0.36744	-0.79	0.4291
y1991	0.07079	0.34572	0.20	0.8379
y1992	-0.47727	0.34452	-1.39	0.1670
y1993	-0.39470	0.34098	-1.16	0.2480
y1994	-0.29206	0.37272	-0.78	0.4339
y1995	-0.62170	0.37354	-1.66	0.0971
y1996	-0.96145	0.38729	-2.48	0.0136
y1997	-1.05577	0.40035	-2.64	0.0088
y1998	-1.25053	0.39057	-3.20	0.0015
y1999	-0.46790	0.35383	-1.32	0.1871
y2000	-0.35840	0.36078	-0.99	0.3213
y2001	-0.52993	0.35768	-1.48	0.1395
y2002	-0.48279	0.36111	-1.34	0.1823
y2003	-0.57225	0.37187	-1.54	0.1249

6 < Age ≤ 12

```
Number of obs =      349
F( 58,    290) =   17.79
Prob > F      =   0.0000
R-squared     =   0.7806
Adj R-squared =   0.7367
Root MSE      =   .34987
```

Source	SS	df	MS
Model	126.32006	58	2.17793
Residual	35.49925	290	0.12241
Total	161.81931	348	

	Coef.	Std. Err.	t	P>|t|
Intercept	-2.32789	0.24046	-9.68	0.0000
Age	0.02163	0.01433	1.51	0.1322
707	0.44096	0.12211	3.61	0.0004
720	0.23085	0.19343	1.19	0.2337
727	(omitted)			
737	0.19680	0.11274	1.75	0.0819
747	1.67007	0.09012	18.53	0.0000
757	1.32843	0.12723	10.44	0.0000
767	1.01565	0.12934	7.85	0.0000
777	(dropped)			
A300	1.83821	0.18700	9.83	0.0000

```
    A310 |  (dropped)
    A319 |  (dropped)
    A320 |  1.13065    0.16419       6.89   0.0000
    A321 |  (dropped)
    A330 |  (dropped)
  BAC111 |  (dropped)
   CV880 |  0.72744    0.22961       3.17   0.0017
   CV990 |  (dropped)
  CVR580 |  (dropped)
    DC10 |  1.37681    0.10200      13.50   0.0000
     DC8 |  0.15447    0.13965       1.11   0.2696
     DC9 | -0.78197    0.14703      -5.32   0.0000
    F100 |  0.14527    0.27794       0.52   0.6016
   L1011 |  0.65859    0.15668       4.20   0.0000
    L188 |  0.66711    0.25487       2.62   0.0093
    MD11 |  1.86912    0.17815      10.49   0.0000
    MD80 |  0.63883    0.13227       4.83   0.0000
    MD90 |  1.26840    0.23922       5.30   0.0000
   SE210 |  0.04662    0.26542       0.18   0.8607
    V700 | -1.16709    0.34613      -3.37   0.0008
   y1965 |  0.18949    0.39660       0.48   0.6332
   y1966 |  0.09353    0.39103       0.24   0.8111
   y1967 |  0.41149    0.32238       1.28   0.2028
   y1968 |  0.31870    0.31228       1.02   0.3083
   y1969 |  0.22374    0.30994       0.72   0.4710
   y1970 |  0.29518    0.28387       1.04   0.2993
   y1971 |  0.02439    0.27260       0.09   0.9288
   y1972 |  0.04213    0.25357       0.17   0.8681
   y1973 |  0.42294    0.23909       1.77   0.0780
   y1974 |  0.37019    0.22805       1.62   0.1056
   y1975 |  0.28922    0.22575       1.28   0.2012
   y1976 |  0.26534    0.22405       1.18   0.2373
   y1977 |  0.30720    0.21580       1.42   0.1557
   y1978 |  0.12454    0.21341       0.58   0.5600
   y1979 | -0.01330    0.20959      -0.05   0.9570
   y1980 |  0.08598    0.21201       0.41   0.6854
   y1981 | -0.00369    0.21143      -0.02   0.9861
   y1982 | -0.28750    0.21407      -1.34   0.1803
   y1983 | -0.07909    0.21209      -0.37   0.7095
   y1984 |  0.01205    0.22982       0.05   0.9582
   y1985 |  (dropped)
   y1986 |  (omitted)
   y1987 |  0.16558    0.22657       0.73   0.4655
   y1988 | -0.02027    0.22260      -0.09   0.9275
   y1989 | -0.09524    0.22761      -0.42   0.6759
   y1990 |  0.00531    0.22363       0.02   0.9811
   y1991 | -0.23208    0.24875      -0.93   0.3516
   y1992 | -0.49060    0.26570      -1.85   0.0659
   y1993 | -0.39215    0.26501      -1.48   0.1400
   y1994 | -0.60860    0.24409      -2.49   0.0132
   y1995 | -0.62285    0.23893      -2.61   0.0096
   y1996 | -0.62456    0.23452      -2.66   0.0082
   y1997 | -0.46757    0.23504      -1.99   0.0476
   y1998 | -0.58169    0.21387      -2.72   0.0069
   y1999 | -0.54871    0.21461      -2.56   0.0111
   y2000 | -0.57475    0.21369      -2.69   0.0076
```

```
      y2001 | -0.34275    0.21386        -1.60   0.1101
      y2002 | -0.29015    0.21541        -1.35   0.1790
      y2003 | -0.40678    0.21865        -1.86   0.0638
-----------------------------------------------------------
```

12 < Age

```
Number of obs =      299
F( 40,   258) =    16.88
Prob > F      =   0.0000
R-squared     =   0.7235
Adj R-squared =   0.6807
Root MSE      =   .43943

     Source |        SS        df        MS
-------------+------------------------------
      Model |   130.38446      40     3.25961
   Residual |    49.81983     258     0.19310
-------------+------------------------------
      Total |   180.20430     298

-----------------------------------------------------------
             |      Coef.   Std. Err.        t    P>|t|
-------------+---------------------------------------------
   Intercept | -1.85617    0.22682        -8.18   0.0000
         Age |  0.00803    0.01277         0.63   0.5302
         707 | -0.73203    0.31340        -2.34   0.0203
         720 |  (dropped)
         727 |  (omitted)
         737 | -0.00197    0.11968        -0.02   0.9869
         747 |  1.20139    0.10813        11.11   0.0000
         757 |  0.97074    0.21119         4.60   0.0000
         767 |  0.10667    0.23563         0.45   0.6512
         777 |  (dropped)
        A300 |  0.79885    0.29505         2.71   0.0072
        A310 |  (dropped)
        A319 |  (dropped)
        A320 |  (dropped)
        A321 |  (dropped)
        A330 |  (dropped)
      BAC111 |  (dropped)
       CV880 | -0.18033    0.46126        -0.39   0.6961
       CV990 |  (dropped)
      CVR580 | -0.05571    0.34799        -0.16   0.8729
        DC10 |  1.24808    0.08206        15.21   0.0000
         DC8 | -0.21807    0.13668        -1.60   0.1118
         DC9 | -0.44970    0.11585        -3.88   0.0001
        F100 |  (dropped)
       L1011 |  0.44026    0.14639         3.01   0.0029
        L188 |  (dropped)
        MD11 |  (dropped)
        MD80 |  0.13563    0.13860         0.98   0.3287
        MD90 |  (dropped)
       SE210 |  (dropped)
        V700 |  (dropped)
```

```
     y1965 |  (dropped)
     y1966 |  (dropped)
     y1967 |  (dropped)
     y1968 | -1.76027     0.46135       -3.82   0.0002
     y1969 |  (dropped)
     y1970 |  (dropped)
     y1971 |  (dropped)
     y1972 |  (dropped)
     y1973 |  (dropped)
     y1974 |  (dropped)
     y1975 |  (dropped)
     y1976 |  (dropped)
     y1977 |  0.44023     0.35588        1.24   0.2172
     y1978 |  0.42137     0.35551        1.19   0.2370
     y1979 |  0.56251     0.31683        1.78   0.0770
     y1980 |  0.29000     0.27388        1.06   0.2906
     y1981 | -0.21975     0.24865       -0.88   0.3776
     y1982 | -0.31631     0.23769       -1.33   0.1844
     y1983 | -0.80448     0.22358       -3.60   0.0004
     y1984 | -0.62220     0.20303       -3.06   0.0024
     y1985 |  (dropped)
     y1986 |  (omitted)
     y1987 | -0.05398     0.17336       -0.31   0.7558
     y1988 | -0.01960     0.17701       -0.11   0.9119
     y1989 |  0.09580     0.18643        0.51   0.6078
     y1990 |  0.02015     0.18776        0.11   0.9146
     y1991 |  0.16520     0.17947        0.92   0.3582
     y1992 |  0.01990     0.18372        0.11   0.9138
     y1993 | -0.08613     0.19205       -0.45   0.6542
     y1994 | -0.14696     0.18678       -0.79   0.4321
     y1995 | -0.20496     0.18981       -1.08   0.2812
     y1996 | -0.04926     0.19352       -0.25   0.7993
     y1997 | -0.04081     0.19241       -0.21   0.8322
     y1998 |  0.11424     0.18855        0.61   0.5451
     y1999 |  0.14453     0.19248        0.75   0.4534
     y2000 |  0.14705     0.19798        0.74   0.4583
     y2001 |  0.22362     0.20047        1.12   0.2657
     y2002 |  0.02142     0.20805        0.10   0.9181
     y2003 |  0.09607     0.20919        0.46   0.6465
---------------------------------------------------------------
```

Burden Cost Regressions

Age ≤ 6

```
Number of obs =      355
F( 63,   294) =     5.86
Prob > F      =   0.0000
R-squared     =   0.5601
Adj R-squared =   0.4646
Root MSE      =   .61115
```

Source	SS	df	MS
Model	137.93020	63	2.18937
Residual	108.31547	290	0.37350
Total	246.24567	353	

	Coef.	Std. Err.	t	P>|t|
Intercept	-3.09879	0.36460	-8.50	0.0000
Age	0.13945	0.02596	5.37	0.0000
707	-0.12753	0.21393	-0.60	0.5516
720	-0.27227	0.30275	-0.90	0.3692
727	(omitted)			
737	0.34745	0.25959	1.34	0.1818
747	0.83141	0.19910	4.18	0.0000
757	0.75902	0.32075	2.37	0.0186
767	0.99723	0.31138	3.20	0.0015
777	0.97060	0.34546	2.81	0.0053
A300	0.89842	0.38838	2.31	0.0214
A310	0.07794	0.43893	0.18	0.8592
A319	-0.17408	0.35860	-0.49	0.6277
A320	0.29958	0.33268	0.90	0.3686
A321	-0.88101	0.48834	-1.80	0.0723
A330	0.25224	0.46157	0.55	0.5852
BAC111	0.43378	0.27916	1.55	0.1213
CV880	0.08669	0.47942	0.18	0.8566
CV990	0.75402	0.39644	1.90	0.0582
CVR580	(dropped)			
DC10	0.52666	0.21955	2.40	0.0171
DC8	0.07260	0.24991	0.29	0.7716
DC9	-0.06796	0.24547	-0.28	0.7821
F100	0.75327	0.39892	1.89	0.0600
L1011	1.09507	0.26593	4.12	0.0000
L188	(dropped)			
MD11	1.29432	0.35047	3.69	0.0003
MD80	0.28821	0.35125	0.82	0.4126
MD90	0.32287	0.38664	0.84	0.4044
SE210	-0.03487	0.47750	-0.07	0.9418
V700	(dropped)			
y1965	1.64005	0.40805	4.02	0.0000
y1966	1.49425	0.39561	3.78	0.0002
y1967	1.36579	0.39838	3.43	0.0007
y1968	1.37803	0.38542	3.58	0.0004
y1969	1.29706	0.38319	3.38	0.0008
y1970	1.33843	0.37109	3.61	0.0004
y1971	1.14043	0.37216	3.06	0.0024
y1972	1.27900	0.37038	3.45	0.0006
y1973	1.14218	0.37380	3.06	0.0025
y1974	1.37663	0.38159	3.61	0.0004
y1975	1.13033	0.39102	2.89	0.0041
y1976	1.16949	0.39156	2.99	0.0031
y1977	1.06688	0.42637	2.50	0.0129
y1978	0.93895	0.44432	2.11	0.0354
y1979	1.01322	0.53371	1.90	0.0586

y1980	0.87681	0.53399	1.64	0.1017
y1981	0.75657	0.69951	1.08	0.2803
y1982	0.75704	0.49725	1.52	0.1290
y1983	0.52083	0.42911	1.21	0.2258
y1984	0.51861	0.35384	1.47	0.1438
y1985	(dropped)			
y1986	(omitted)			
y1987	0.12334	0.30886	0.40	0.6899
y1988	0.01105	0.30946	0.04	0.9715
y1989	-0.11443	0.29590	-0.39	0.6993
y1990	-0.02136	0.30497	-0.07	0.9442
y1991	0.14023	0.28695	0.49	0.6254
y1992	0.11369	0.28595	0.40	0.6912
y1993	0.22952	0.28301	0.81	0.4180
y1994	0.47800	0.30935	1.55	0.1234
y1995	0.23105	0.31004	0.75	0.4567
y1996	0.21940	0.32145	0.68	0.4955
y1997	0.36606	0.33229	1.10	0.2715
y1998	-0.06404	0.32417	-0.20	0.8435
y1999	-0.07323	0.29368	-0.25	0.8033
y2000	0.40536	0.29944	1.35	0.1769
y2001	0.36304	0.29687	1.22	0.2224
y2002	0.55343	0.29972	1.85	0.0658
y2003	0.57407	0.30865	1.86	0.0639

6 < Age ≤ 12

Number of obs	=	349
F(58, 290)	=	4.94
Prob > F	=	0.0000
R-squared	=	0.4968
Adj R-squared	=	0.3961
Root MSE	=	.43774

Source	SS	df	MS
Model	54.85849	58	0.94584
Residual	55.56858	290	0.19162
Total	110.42708	348	

	Coef.	Std. Err.	t	P>|t|
Intercept	-2.35466	0.30085	-7.83	0.0000
Age	0.06668	0.01792	3.72	0.0002
707	0.18242	0.15278	1.19	0.2335
720	-0.21754	0.24200	-0.90	0.3694
727	(omitted)			
737	0.06684	0.14105	0.47	0.6360
747	0.88591	0.11275	7.86	0.0000
757	0.37102	0.15918	2.33	0.0204
767	0.68353	0.16182	4.22	0.0000

```
     777 |  (dropped)
    A300 |  0.98731     0.23396        4.22   0.0000
    A310 |  (dropped)
    A319 |  (dropped)
    A320 | -0.25783     0.20542       -1.26   0.2105
    A321 |  (dropped)
    A330 |  (dropped)
  BAC111 |  (dropped)
   CV880 |  0.24551     0.28727        0.85   0.3935
   CV990 |  (dropped)
  CVR580 |  (dropped)
    DC10 |  0.78436     0.12762        6.15   0.0000
     DC8 |  0.21782     0.17472        1.25   0.2135
     DC9 | -0.04696     0.18395       -0.26   0.7987
    F100 |  0.17404     0.34775        0.50   0.6171
   L1011 |  1.12028     0.19603        5.71   0.0000
    L188 | -0.53038     0.31888       -1.66   0.0973
    MD11 |  0.73558     0.22289        3.30   0.0011
    MD80 |  0.19708     0.16549        1.19   0.2347
    MD90 |  0.41736     0.29930        1.39   0.1642
   SE210 |  0.04621     0.33208        0.14   0.8894
    V700 | -0.53244     0.43305       -1.23   0.2199
   y1965 |  1.01488     0.49621        2.05   0.0417
   y1966 |  1.02737     0.48923        2.10   0.0366
   y1967 |  1.08990     0.40334        2.70   0.0073
   y1968 |  0.90008     0.39071        2.30   0.0219
   y1969 |  0.85569     0.38778        2.21   0.0281
   y1970 |  0.85477     0.35516        2.41   0.0167
   y1971 |  0.83164     0.34106        2.44   0.0154
   y1972 |  0.57143     0.31725        1.80   0.0727
   y1973 |  0.54639     0.29913        1.83   0.0688
   y1974 |  0.61615     0.28532        2.16   0.0316
   y1975 |  0.53792     0.28244        1.90   0.0578
   y1976 |  0.50650     0.28031        1.81   0.0718
   y1977 |  0.46125     0.27000        1.71   0.0886
   y1978 |  0.46859     0.26701        1.75   0.0803
   y1979 |  0.34495     0.26222        1.32   0.1894
   y1980 |  0.34940     0.26526        1.32   0.1888
   y1981 |  0.37299     0.26453        1.41   0.1596
   y1982 |  0.17515     0.26783        0.65   0.5137
   y1983 |  0.09576     0.26535        0.36   0.7185
   y1984 | -0.14815     0.28754       -0.52   0.6068
   y1985 |  (dropped)
   y1986 |  (omitted)
   y1987 | -0.39691     0.28347       -1.40   0.1625
   y1988 | -0.23692     0.27851       -0.85   0.3956
   y1989 | -0.15601     0.28477       -0.55   0.5842
   y1990 | -0.15964     0.27979       -0.57   0.5687
   y1991 |  0.05163     0.31123        0.17   0.8684
   y1992 |  0.13639     0.33243        0.41   0.6819
   y1993 |  0.09149     0.33157        0.28   0.7828
   y1994 | -0.07006     0.30539       -0.23   0.8187
   y1995 | -0.17089     0.29893       -0.57   0.5680
   y1996 | -0.12828     0.29342       -0.44   0.6623
   y1997 | -0.14592     0.29406       -0.50   0.6201
   y1998 | -0.23155     0.26758       -0.87   0.3876
```

y1999	-0.14112	0.26851	-0.53	0.5996
y2000	-0.14171	0.26735	-0.53	0.5965
y2001	0.20872	0.26757	0.78	0.4360
y2002	0.09323	0.26951	0.35	0.7297
y2003	0.02538	0.27356	0.09	0.9261

12 < Age

```
Number of obs =      299
F( 40,   258) =     5.93
Prob > F      =   0.0000
R-squared     =   0.4791
Adj R-squared =   0.3984
Root MSE      =   .42798
```

Source	SS	df	MS
Model	43.47078	40	1.08677
Residual	47.25600	258	0.18316
Total	90.72678	298	

	Coef.	Std. Err.	t	P>|t|
Intercept	-1.52821	0.22091	-6.92	0.0000
Age	0.00739	0.01244	0.59	0.5528
707	0.13694	0.30523	0.45	0.6541
720	(dropped)			
727	(omitted)			
737	-0.02934	0.11656	-0.25	0.8015
747	0.80118	0.10531	7.61	0.0000
757	-0.70369	0.20569	-3.42	0.0007
767	0.42430	0.22949	1.85	0.0656
777	(dropped)			
A300	0.97902	0.28736	3.41	0.0008
A310	(dropped)			
A319	(dropped)			
A320	(dropped)			
A321	(dropped)			
A330	(dropped)			
BAC111	(dropped)			
CV880	0.64293	0.44923	1.43	0.1536
CV990	(dropped)			
CVR580	-0.28056	0.33892	-0.83	0.4085
DC10	0.56152	0.07992	7.03	0.0000
DC8	0.41244	0.13312	3.10	0.0022
DC9	-0.23474	0.11283	-2.08	0.0385
F100	(dropped)			
L1011	0.51530	0.14258	3.61	0.0004
L188	(dropped)			
MD11	(dropped)			
MD80	0.06939	0.13499	0.51	0.6076
MD90	(dropped)			

```
   SE210 |   (dropped)
    V700 |   (dropped)
   y1965 |   (dropped)
   y1966 |   (dropped)
   y1967 |   (dropped)
   y1968 |   0.18051     0.44933         0.40    0.6882
   y1969 |   (dropped)
   y1970 |   (dropped)
   y1971 |   (dropped)
   y1972 |   (dropped)
   y1973 |   (dropped)
   y1974 |   (dropped)
   y1975 |   (dropped)
   y1976 |   (dropped)
   y1977 |   0.43263     0.34660         1.25    0.2131
   y1978 |   0.33681     0.34625         0.97    0.3316
   y1979 |   0.47273     0.30857         1.53    0.1267
   y1980 |   0.52211     0.26674         1.96    0.0514
   y1981 |   0.50429     0.24217         2.08    0.0383
   y1982 |   0.38768     0.23150         1.67    0.0952
   y1983 |   0.29078     0.21775         1.34    0.1829
   y1984 |   0.15406     0.19774         0.78    0.4366
   y1985 |   (dropped)
   y1986 |   (omitted)
   y1987 |   0.02227     0.16884         0.13    0.8952
   y1988 |  -0.02489     0.17239        -0.14    0.8853
   y1989 |   0.12709     0.18157         0.70    0.4846
   y1990 |   0.22991     0.18287         1.26    0.2098
   y1991 |   0.25572     0.17479         1.46    0.1447
   y1992 |   0.23668     0.17893         1.32    0.1871
   y1993 |   0.26415     0.18704         1.41    0.1591
   y1994 |   0.24101     0.18192         1.32    0.1864
   y1995 |   0.29638     0.18486         1.60    0.1101
   y1996 |   0.27737     0.18847         1.47    0.1423
   y1997 |   0.22237     0.18740         1.19    0.2365
   y1998 |   0.24511     0.18364         1.33    0.1831
   y1999 |   0.22481     0.18747         1.20    0.2316
   y2000 |   0.22265     0.19281         1.15    0.2493
   y2001 |   0.20275     0.19524         1.04    0.3000
   y2002 |   0.08018     0.20263         0.40    0.6927
   y2003 |   0.24673     0.20374         1.21    0.2270
-------------------------------------------------------
```

Next, RAND tested whether it would be valuable to include airline dummy variables in the total maintenance cost per flying hour regressions. Some of the airline coefficients are statistically significant, but the Age coefficients (of central interest) do not meaningfully change relative to the baseline regressions.

Airline Dummy Variable Regressions

Age ≤ 6

```
Number of obs =      355
F( 69,    284) =    14.38
Prob > F       =   0.0000
R-squared      =   0.7775
Adj R-squared =   0.7234
Root MSE       =   .37592
```

Source	SS	df	MS
Model	140.23219	69	2.03235
Residual	40.13440	284	0.14132
Total	180.36659	353	

	Coef.	Std. Err.	t	P>|t|
Intercept	-1.76524	0.23136	-7.63	0.0000
Age	0.16579	0.01634	10.15	0.0000
707	-0.02747	0.13381	-0.21	0.8375
720	-0.33859	0.18688	-1.81	0.0711
727	(omitted)			
737	-0.07288	0.16402	-0.44	0.6571
747	1.26472	0.12576	10.06	0.0000
757	0.62572	0.20424	3.06	0.0024
767	0.73537	0.19767	3.72	0.0002
777	1.07965	0.22121	4.88	0.0000
A300	1.09081	0.24489	4.45	0.0000
A310	0.68103	0.27397	2.49	0.0135
A319	-0.05267	0.24765	-0.21	0.8317
A320	0.51200	0.22342	2.29	0.0227
A321	-0.53369	0.33787	-1.58	0.1153
A330	1.03157	0.31332	3.29	0.0011
BAC111	0.11561	0.17710	0.65	0.5144
CV880	-0.15851	0.30232	-0.52	0.6005
CV990	0.35150	0.24831	1.42	0.1580
CVR580	(dropped)			
DC10	0.96104	0.14050	6.84	0.0000
DC8	-0.40518	0.15787	-2.57	0.0108
DC9	-0.31945	0.15959	-2.00	0.0463
F100	0.72609	0.24983	2.91	0.0039
L1011	1.17061	0.16477	7.10	0.0000
L188	(dropped)			
MD11	1.24509	0.21792	5.71	0.0000
MD80	0.00228	0.21936	0.01	0.9917
MD90	0.56147	0.24315	2.31	0.0217
SE210	-0.55890	0.29580	-1.89	0.0598
V700	(dropped)			
y1965	1.37317	0.25679	5.35	0.0000
y1966	1.18611	0.24895	4.76	0.0000

```
        y1967 |  1.16060    0.25020       4.64   0.0000
        y1968 |  1.06992    0.24245       4.41   0.0000
        y1969 |  0.96592    0.24092       4.01   0.0000
        y1970 |  0.84051    0.23253       3.61   0.0004
        y1971 |  0.64910    0.23293       2.79   0.0057
        y1972 |  0.61502    0.23031       2.67   0.0080
        y1973 |  0.50134    0.23204       2.16   0.0316
        y1974 |  0.63138    0.23673       2.67   0.0081
        y1975 |  0.51509    0.24213       2.13   0.0343
        y1976 |  0.41122    0.24284       1.69   0.0915
        y1977 |  0.36175    0.26375       1.37   0.1713
        y1978 |  0.20454    0.27526       0.74   0.4581
        y1979 |  0.17799    0.33009       0.54   0.5902
        y1980 |  0.08986    0.33029       0.27   0.7858
        y1981 | -0.04421    0.43106      -0.10   0.9184
        y1982 |  0.04876    0.30692       0.16   0.8739
        y1983 | -0.09881    0.26446      -0.37   0.7090
        y1984 | -0.08310    0.21826      -0.38   0.7037
        y1985 |  (dropped)
        y1986 |  (omitted)
        y1987 |  0.06566    0.19038       0.34   0.7304
        y1988 | -0.07922    0.19122      -0.41   0.6790
        y1989 | -0.13534    0.18317      -0.74   0.4606
        y1990 | -0.07111    0.18875      -0.38   0.7066
        y1991 |  0.07567    0.17783       0.43   0.6708
        y1992 | -0.15058    0.17734      -0.85   0.3966
        y1993 | -0.16051    0.17591      -0.91   0.3623
        y1994 | -0.07179    0.19263      -0.37   0.7097
        y1995 | -0.24200    0.19349      -1.25   0.2121
        y1996 | -0.32506    0.20082      -1.62   0.1066
        y1997 | -0.28738    0.20776      -1.38   0.1677
        y1998 | -0.46886    0.20993      -2.23   0.0263
        y1999 | -0.05167    0.19268      -0.27   0.7888
        y2000 |  0.01980    0.19599       0.10   0.9196
        y2001 | -0.09360    0.19610      -0.48   0.6335
        y2002 |  0.01700    0.19841       0.09   0.9318
        y2003 |  0.06366    0.20216       0.31   0.7531
       Alaska |  (dropped)
 America West |  0.03491    0.18822       0.19   0.8530
     American |  (omitted)
  Continental | -0.17485    0.12903      -1.36   0.1765
        Delta |  0.04852    0.07201       0.67   0.5010
    Northwest | -0.40281    0.07870      -5.12   0.0000
    Southwest |  (dropped)
       United |  0.16428    0.07039       2.33   0.0203
   US Airways | -0.15895    0.14735      -1.08   0.2816
-----------------------------------------------------
```

6 < Age ≤ 12

```
Number of obs =      349
F( 66,    282) =   28.10
Prob > F      =  0.0000
R-squared     =  0.8680
Adj R-squared =  0.8371
```

Root MSE = .17357

Source	SS	df	MS
Model	55.86379	66	0.84642
Residual	8.49541	282	0.03013
Total	64.35920	348	

	Coef.	Std. Err.	t	P>|t|
Intercept	-0.96468	0.12855	-7.50	0.0000
Age	0.03287	0.00767	4.28	0.0000
707	0.28568	0.06198	4.61	0.0000
720	0.14926	0.09784	1.53	0.1283
727	(omitted)			
737	0.16433	0.05723	2.87	0.0044
747	1.22812	0.04617	26.60	0.0000
757	0.57933	0.06746	8.59	0.0000
767	0.61643	0.06711	9.18	0.0000
777	(dropped)			
A300	1.14034	0.09722	11.73	0.0000
A310	(dropped)			
A319	(dropped)			
A320	0.46127	0.09920	4.65	0.0000
A321	(dropped)			
A330	(dropped)			
BAC111	(dropped)			
CV880	0.60490	0.12038	5.02	0.0000
CV990	(dropped)			
CVR580	(dropped)			
DC10	0.93171	0.05139	18.13	0.0000
DC8	0.26518	0.07231	3.67	0.0003
DC9	-0.07218	0.07842	-0.92	0.3581
F100	-0.05129	0.14219	-0.36	0.7186
L1011	0.87759	0.07917	11.09	0.0000
L188	0.28723	0.13037	2.20	0.0284
MD11	0.94791	0.09165	10.34	0.0000
MD80	0.25399	0.06685	3.80	0.0002
MD90	0.67592	0.12208	5.54	0.0000
SE210	0.13577	0.13348	1.02	0.3100
V700	-0.29459	0.17540	-1.68	0.0941
y1965	0.32469	0.20253	1.60	0.1100
y1966	0.35474	0.19924	1.78	0.0761
y1967	0.49518	0.16537	2.99	0.0030
y1968	0.42384	0.15952	2.66	0.0083
y1969	0.39619	0.15746	2.52	0.0124
y1970	0.46129	0.14429	3.20	0.0015
y1971	0.28364	0.13849	2.05	0.0415
y1972	0.19350	0.12859	1.50	0.1335
y1973	0.34046	0.12105	2.81	0.0053
y1974	0.35207	0.11550	3.05	0.0025
y1975	0.30141	0.11402	2.64	0.0087
y1976	0.28525	0.11288	2.53	0.0120
y1977	0.27049	0.10848	2.49	0.0132

```
       y1978 |  0.17211     0.10736        1.60    0.1100
       y1979 |  0.09678     0.10516        0.92    0.3582
       y1980 |  0.15273     0.10618        1.44    0.1514
       y1981 |  0.13347     0.10556        1.26    0.2071
       y1982 | -0.03926     0.10667       -0.37    0.7131
       y1983 |  0.02635     0.10561        0.25    0.8032
       y1984 | -0.01163     0.11433       -0.10    0.9190
       y1985 |  (dropped)
       y1986 |  (omitted)
       y1987 |  0.07079     0.11254        0.63    0.5299
       y1988 |  0.02688     0.11082        0.24    0.8085
       y1989 |  0.08122     0.11362        0.71    0.4753
       y1990 |  0.13159     0.11147        1.18    0.2388
       y1991 |  0.13389     0.12387        1.08    0.2807
       y1992 | -0.04342     0.13220       -0.33    0.7428
       y1993 | -0.04433     0.13176       -0.34    0.7368
       y1994 | -0.23251     0.12187       -1.91    0.0574
       y1995 | -0.22597     0.11930       -1.89    0.0592
       y1996 | -0.22670     0.11713       -1.94    0.0539
       y1997 | -0.18455     0.11761       -1.57    0.1177
       y1998 | -0.18852     0.10795       -1.75    0.0818
       y1999 | -0.15161     0.10814       -1.40    0.1620
       y2000 | -0.02337     0.10788       -0.22    0.8287
       y2001 |  0.11433     0.10799        1.06    0.2906
       y2002 |  0.09418     0.10907        0.86    0.3886
       y2003 | -0.05441     0.11132       -0.49    0.6254
      Alaska |  0.06502     0.06995        0.93    0.3534
America West | -0.16807     0.10071       -1.67    0.0963
    American |  (omitted)
 Continental |  0.00169     0.09033        0.02    0.9851
       Delta | -0.16655     0.03601       -4.63    0.0000
   Northwest | -0.22779     0.03154       -7.22    0.0000
   Southwest | -0.11084     0.08891       -1.25    0.2135
      United | -0.02248     0.03119       -0.72    0.4717
  US Airways | -0.22831     0.07041       -3.24    0.0013
--------------------------------------------------------
```

12 < Age

```
Number of obs =      299
F( 47,    251) =   19.03
Prob > F      =   0.0000
R-squared     =   0.7809
Adj R-squared =   0.7398
Root MSE      =   .23141

      Source |        SS          df         MS
-------------+------------------------------------
       Model |     47.89311      47      1.01900
    Residual |     13.44123     251      0.05355
-------------+------------------------------------
       Total |     61.33434     298
```

	Coef.	Std. Err.	t	P>|t|
Intercept	-0.40240	0.13132	-3.06	0.0024
Age	0.00423	0.00717	0.59	0.5552
707	-0.08629	0.16794	-0.51	0.6078
720	(dropped)			
727	(omitted)			
737	-0.09205	0.07050	-1.31	0.1928
747	0.81997	0.06006	13.65	0.0000
757	0.15082	0.13147	1.15	0.2524
767	0.02995	0.13323	0.22	0.8223
777	(dropped)			
A300	0.68617	0.16269	4.22	0.0000
A310	(dropped)			
A319	(dropped)			
A320	(dropped)			
A321	(dropped)			
A330	(dropped)			
BAC111	(dropped)			
CV880	0.29336	0.24522	1.20	0.2327
CV990	(dropped)			
CVR580	0.06127	0.18899	0.32	0.7461
DC10	0.78516	0.04377	17.94	0.0000
DC8	0.25856	0.07566	3.42	0.0007
DC9	-0.16838	0.06556	-2.57	0.0108
F100	(dropped)			
L1011	0.52458	0.08399	6.25	0.0000
L188	(dropped)			
MD11	(dropped)			
MD80	-0.10041	0.07876	-1.27	0.2035
MD90	(dropped)			
SE210	(dropped)			
V700	(dropped)			
y1965	(dropped)			
y1966	(dropped)			
y1967	(dropped)			
y1968	-0.36819	0.24448	-1.51	0.1333
y1969	(dropped)			
y1970	(dropped)			
y1971	(dropped)			
y1972	(dropped)			
y1973	(dropped)			
y1974	(dropped)			
y1975	(dropped)			
y1976	(dropped)			
y1977	0.18935	0.18769	1.01	0.3140
y1978	0.12645	0.18749	0.67	0.5006
y1979	0.23841	0.16707	1.43	0.1548
y1980	0.24014	0.14455	1.66	0.0979
y1981	0.15178	0.13138	1.16	0.2491
y1982	0.04213	0.12577	0.34	0.7379
y1983	-0.09389	0.11836	-0.79	0.4284
y1984	-0.16462	0.10738	-1.53	0.1265
y1985	(dropped)			
y1986	(omitted)			

```
     y1987 | -0.01280   0.09132     -0.14   0.8886
     y1988 | -0.00314   0.09349     -0.03   0.9733
     y1989 |  0.10877   0.09842      1.11   0.2701
     y1990 |  0.14972   0.09926      1.51   0.1327
     y1991 |  0.21029   0.09483      2.22   0.0275
     y1992 |  0.16208   0.09725      1.67   0.0968
     y1993 |  0.07505   0.10164      0.74   0.4610
     y1994 |  0.08113   0.09902      0.82   0.4134
     y1995 |  0.09487   0.10081      0.94   0.3476
     y1996 |  0.14095   0.10300      1.37   0.1724
     y1997 |  0.15971   0.10281      1.55   0.1216
     y1998 |  0.24163   0.10147      2.38   0.0180
     y1999 |  0.20877   0.10380      2.01   0.0454
     y2000 |  0.25522   0.10662      2.39   0.0174
     y2001 |  0.28002   0.10854      2.58   0.0105
     y2002 |  0.21460   0.11247      1.91   0.0575
     y2003 |  0.27695   0.11448      2.42   0.0163
    Alaska |  0.05977   0.24861      0.24   0.8102
America West |  0.17344   0.10486      1.65   0.0994
  American |  (omitted)
Continental |  0.18447   0.07878      2.34   0.0200
     Delta | -0.09523   0.05679     -1.68   0.0948
 Northwest | -0.17058   0.04715     -3.62   0.0004
 Southwest |  (dropped)
    United |  0.00929   0.04658      0.20   0.8421
US Airways |  0.14554   0.08239      1.77   0.0785
-------------------------------------------------------
```

Next, the estimation was run with both Age*Type and Age*Airline interaction variables included. Some of these age-interaction variables are statistically significant, but no clear pattern was found of either certain types of aircraft aging worse or better than other types of aircraft or certain airlines experiencing worse or better aging of aircraft than other airlines. This "nonresult" is encouraging with respect to possibly generalizing the findings to the Air Force.

Age*Type and Age*Airline Interaction Regressions

Age ≤ 6

```
Number of obs =      355
F(100,    253) =   13.28
Prob > F      =   0.0000
R-squared     =   0.8400
Adj R-squared =   0.7767
Root MSE      =   .33778
```

Source	SS	df	MS
Model	151.50043	100	1.51500
Residual	28.86616	253	0.11410
Total	180.36659	353	

	Coef.	Std. Err.	t	P>|t|
Intercept	-1.50313	0.36503	-4.12	0.0000
Age	0.12530	0.07402	1.69	0.0917
707	0.28710	0.34411	0.83	0.4049
720	-0.18149	0.97733	-0.19	0.8528
727	(omitted)			
737	-0.33067	0.32170	-1.03	0.3050
747	1.03971	0.34905	2.98	0.0032
757	0.00408	0.37694	0.01	0.9914
767	0.59796	0.39133	1.53	0.1278
777	0.51228	0.36809	1.39	0.1652
A300	0.17187	0.44670	0.38	0.7007
A310	1.45797	1.34917	1.08	0.2809
A319	-1.37323	0.40595	-3.38	0.0008
A320	-0.01770	0.39286	-0.05	0.9641
A321	-1.87419	0.67488	-2.78	0.0059
A330	0.74196	0.48458	1.53	0.1270
BAC111	-0.03682	0.33771	-0.11	0.9133
CV880	-0.89495	2.57725	-0.35	0.7287
CV990	-0.23668	1.25779	-0.19	0.8509
CVR580	(dropped)			
DC10	0.45774	0.36576	1.25	0.2119
DC8	0.09012	0.79153	0.11	0.9094
DC9	-0.80968	0.31932	-2.54	0.0118
F100	0.61510	0.42401	1.45	0.1481
L1011	0.43157	0.43483	0.99	0.3219
L188	(dropped)			
MD11	0.73605	0.40167	1.83	0.0681
MD80	-0.99920	0.41896	-2.38	0.0178
MD90	-1.27257	0.46313	-2.75	0.0064
SE210	0.37756	2.39506	0.16	0.8749
V700	(dropped)			
y1965	1.09622	0.31541	3.48	0.0006
y1966	0.95727	0.29425	3.25	0.0013
y1967	0.92141	0.28227	3.26	0.0012
y1968	0.88644	0.26252	3.38	0.0008
y1969	0.83151	0.25341	3.28	0.0012
y1970	0.74936	0.23620	3.17	0.0017
y1971	0.59880	0.23429	2.56	0.0112
y1972	0.64379	0.23637	2.72	0.0069
y1973	0.52041	0.23888	2.18	0.0303
y1974	0.62644	0.25644	2.44	0.0153
y1975	0.51145	0.26274	1.95	0.0527
y1976	0.46572	0.26795	1.74	0.0834
y1977	0.30960	0.27897	1.11	0.2682
y1978	0.19430	0.28297	0.69	0.4929
y1979	0.27306	0.31111	0.88	0.3809

```
       y1980 |  0.24625    0.31350       0.79    0.4329
       y1981 | -0.02977    0.39754      -0.07    0.9404
       y1982 | -0.13100    0.30230      -0.43    0.6651
       y1983 | -0.40014    0.27084      -1.84    0.1408
       y1984 | -0.27940    0.20919      -1.34    0.1829
       y1985 |  (dropped)
       y1986 |  (omitted)
       y1987 |  0.15574    0.17581       0.89    0.3765
       y1988 |  0.12213    0.18129       0.67    0.5011
       y1989 |  0.07979    0.17660       0.45    0.6518
       y1990 |  0.09628    0.18152       0.53    0.5963
       y1991 |  0.18760    0.17730       1.06    0.2910
       y1992 |  0.02092    0.17965       0.12    0.9074
       y1993 | -0.04364    0.17824      -0.24    0.8068
       y1994 | -0.03651    0.19182      -0.19    0.8492
       y1995 | -0.05658    0.19370      -0.29    0.7705
       y1996 | -0.15749    0.20353      -0.77    0.4398
       y1997 | -0.08482    0.21569      -0.39    0.6945
       y1998 | -0.30167    0.21833      -1.38    0.1683
       y1999 |  0.09912    0.19563       0.51    0.6128
       y2000 |  0.05387    0.19730       0.27    0.7850
       y2001 | -0.08313    0.19499      -0.43    0.6702
       y2002 | -0.01872    0.19707      -0.09    0.9244
       y2003 | -0.01161    0.20707      -0.06    0.9553
     Age*707 | -0.07138    0.09477      -0.75    0.4520
     Age*720 |  0.01686    0.19570       0.09    0.9314
     Age*727 |  (omitted)
     Age*737 |  0.03494    0.08308       0.42    0.6744
     Age*747 |  0.03369    0.09526       0.35    0.7239
     Age*757 |  0.10559    0.08536       1.24    0.2172
     Age*767 |  0.01107    0.09087       0.12    0.9031
     Age*777 |  0.14244    0.09431       1.51    0.1322
    Age*A300 |  0.23103    0.11683       1.98    0.0491
    Age*A310 | -0.15288    0.28513      -0.54    0.5923
    Age*A319 |  0.53244    0.12002       4.44    0.0000
    Age*A320 |  0.11586    0.09373       1.24    0.2176
    Age*A321 |  0.55176    0.27457       2.01    0.0455
    Age*A330 |  0.04790    0.19388       0.25    0.8051
  Age*BAC111 |  0.05506    0.09739       0.57    0.5723
   Age*CV880 |  0.24329    0.50556       0.48    0.6308
   Age*CV990 |  0.16832    0.27951       0.60    0.5476
  Age*CVR580 |  (dropped)
    Age*DC10 |  0.11401    0.09802       1.16    0.2459
     Age*DC8 | -0.02997    0.16821      -0.18    0.8587
     Age*DC9 |  0.13099    0.09705       1.35    0.1783
    Age*F100 | -0.08662    0.11471      -0.76    0.4509
   Age*L1011 |  0.19106    0.11061       1.73    0.0853
    Age*L188 |  (dropped)
    Age*MD11 |  0.06616    0.09537       0.69    0.4885
    Age*MD80 |  0.20287    0.09609       2.11    0.0357
    Age*MD90 |  0.50465    0.12090       4.17    0.0000
   Age*SE210 | -0.11743    0.50533      -0.23    0.8164
    Age*V700 |  (dropped)
      Alaska |  (dropped)
America West |  0.88194    0.33088       2.67    0.0082
    American |  (omitted)
```

	Coef.	Std. Err.	t	P>|t|
Continental	-0.48169	0.23180	-2.08	0.0387
Delta	0.58051	0.15173	3.83	0.0002
Northwest	-0.43757	0.16099	-2.72	0.0070
Southwest	(dropped)			
United	0.45144	0.13378	3.37	0.0009
US Airways	0.60135	0.27612	2.18	0.0303
Age*Alaska	(dropped)			
Age*America West	-0.25157	0.09011	-2.79	0.0056
Age*American	(omitted)			
Age*Continental	0.11175	0.06997	1.60	0.1115
Age*Delta	-0.16445	0.04170	-3.94	0.0001
Age*Northwest	0.01868	0.04515	0.41	0.6794
Age*Southwest	(dropped)			
Age*United	-0.10272	0.03794	-2.71	0.0072
Age*US Airways	-0.33548	0.10620	-3.16	0.0018

6 < Age ≤ 12

Number of obs = 349
F(94, 254) = 22.46
Prob > F = 0.0000
R-squared = 0.8926
Adj R-squared = 0.8529
Root MSE = .16494

Source	SS	df	MS
Model	57.44874	94	0.61116
Residual	6.91046	254	0.02721
Total	64.35920	348	

	Coef.	Std. Err.	t	P>|t|
Intercept	-0.86917	0.23679	-3.67	0.0003
Age	0.02766	0.02315	1.19	0.2334
707	0.60928	0.29117	2.09	0.0374
720	0.59296	0.64274	0.92	0.3571
727	(omitted)			
737	-0.03187	0.25403	-0.13	0.9003
747	0.86920	0.24214	3.59	0.0004
757	0.25488	0.24519	1.04	0.3015
767	0.47200	0.27579	1.71	0.0882
777	(dropped)			
A300	1.49689	0.43337	3.45	0.0006
A310	(dropped)			
A319	(dropped)			
A320	-0.29717	0.49957	-0.59	0.5525
A321	(dropped)			
A330	(dropped)			
BAC111	(dropped)			
CV880	1.13734	0.71675	1.59	0.1138
CV990	(dropped)			

CVR580	(dropped)			
DC10	1.18439	0.27286	4.34	0.0000
DC8	0.14550	0.39213	0.37	0.7109
DC9	-0.27642	0.38827	-0.71	0.4772
F100	5.41096	2.27217	2.38	0.0180
L1011	0.53050	0.38064	1.39	0.1646
L188	-1.42729	1.00695	-1.42	0.1576
MD11	2.67249	0.59908	4.46	0.0000
MD80	0.13041	0.28936	0.45	0.6526
MD90	1.03438	0.90146	1.15	0.2523
SE210	-0.13359	0.89919	-0.15	0.8820
V700	-1.51455	2.07545	-0.73	0.4662
y1965	0.66957	0.47593	1.41	0.1607
y1966	0.53660	0.38811	1.38	0.1680
y1967	0.34439	0.29174	1.18	0.2389
y1968	0.20169	0.24793	0.81	0.4167
y1969	0.16778	0.20957	0.80	0.4241
y1970	0.25622	0.18040	1.42	0.1567
y1971	0.17789	0.16112	1.10	0.2706
y1972	0.10647	0.14354	0.74	0.4589
y1973	0.25434	0.12729	2.00	0.0468
y1974	0.28608	0.11798	2.42	0.0160
y1975	0.24946	0.11455	2.18	0.0303
y1976	0.24746	0.11280	2.19	0.0292
y1977	0.27185	0.10985	2.47	0.0140
y1978	0.15272	0.10960	1.39	0.1647
y1979	0.05144	0.10959	0.47	0.6392
y1980	0.10776	0.10917	0.99	0.3246
y1981	0.08030	0.10569	0.76	0.4481
y1982	-0.07370	0.10384	-0.71	0.4785
y1983	0.00753	0.10191	0.07	0.9411
y1984	-0.01662	0.11060	-0.15	0.8807
y1985	(dropped)			
y1986	(omitted)			
y1987	0.05520	0.10840	0.51	0.6110
y1988	-0.00655	0.10866	-0.06	0.9520
y1989	0.01765	0.11468	0.15	0.8778
y1990	0.05842	0.11342	0.52	0.6069
y1991	0.06122	0.12301	0.50	0.6192
y1992	-0.06823	0.13051	-0.52	0.6015
y1993	-0.08294	0.13050	-0.64	0.5256
y1994	-0.27496	0.12094	-2.27	0.0238
y1995	-0.28534	0.11858	-2.41	0.0168
y1996	-0.27509	0.11589	-2.37	0.0184
y1997	-0.21609	0.11672	-1.85	0.0653
y1998	-0.22765	0.10833	-2.10	0.0366
y1999	-0.19844	0.10907	-1.82	0.0700
y2000	-0.07897	0.10926	-0.72	0.4705
y2001	0.06278	0.10941	0.57	0.5666
y2002	0.02551	0.11069	0.23	0.8179
y2003	-0.07697	0.11484	-0.67	0.5033
Age*707	-0.03531	0.03255	-1.08	0.2790
Age*720	-0.03500	0.06241	-0.56	0.5753
Age*727	(omitted)			
Age*737	0.02315	0.02778	0.83	0.4055
Age*747	0.03887	0.02651	1.47	0.1438

```
         Age*757 |  0.03772    0.02722        1.39   0.1670
         Age*767 |  0.01670    0.03024        0.55   0.5812
         Age*777 |  (dropped)
        Age*A300 | -0.04158    0.04815       -0.86   0.3886
        Age*A310 |  (dropped)
        Age*A319 |  (dropped)
        Age*A320 |  0.10435    0.06424        1.62   0.1055
        Age*A321 |  (dropped)
        Age*A330 |  (dropped)
      Age*BAC111 |  (dropped)
       Age*CV880 | -0.04434    0.06951       -0.64   0.5241
       Age*CV990 |  (dropped)
      Age*CVR580 |  (dropped)
        Age*DC10 | -0.03234    0.03100       -1.04   0.2979
         Age*DC8 |  0.01926    0.03975        0.48   0.6285
         Age*DC9 |  0.02243    0.04322        0.52   0.6042
        Age*F100 | -0.51652    0.21522       -2.40   0.0171
       Age*L1011 |  0.03788    0.04300        0.88   0.3792
        Age*L188 |  0.20652    0.09700        2.13   0.0342
        Age*MD11 | -0.21224    0.07212       -2.94   0.0036
        Age*MD80 |  0.01316    0.03140        0.42   0.6756
        Age*MD90 | -0.05553    0.12307       -0.45   0.6523
       Age*SE210 |  0.04997    0.10368        0.48   0.6302
        Age*V700 |  0.10773    0.18411        0.59   0.5590
          Alaska | -0.06039    0.33234       -0.18   0.8560
    America West |  0.02068    0.41027        0.05   0.9598
        American |  (omitted)
     Continental | -0.51888    0.86760       -0.60   0.5503
           Delta |  0.03460    0.19965        0.17   0.8625
       Northwest | -0.35907    0.16979       -2.11   0.0354
       Southwest | -0.24478    0.85621       -0.29   0.7752
          United |  0.23115    0.17621        1.31   0.1908
      US Airways |  0.04159    0.50603        0.08   0.9346
      Age*Alaska |  0.01612    0.03707        0.44   0.6639
Age*America West | -0.02701    0.04643       -0.58   0.5613
    Age*American |  (omitted)
 Age*Continental |  0.08199    0.12803        0.64   0.5225
       Age*Delta | -0.02341    0.02256       -1.04   0.3004
   Age*Northwest |  0.01290    0.01921        0.67   0.5023
   Age*Southwest |  0.01460    0.09829        0.15   0.8820
      Age*United | -0.02900    0.01967       -1.47   0.1417
  Age*US Airways | -0.02975    0.04922       -0.60   0.5462
-------------------------------------------------------------
```

12 < Age

```
Number of obs =     299
F( 65,   233) =   16.51
Prob > F      =  0.0000
R-squared     =  0.8217
Adj R-squared =  0.7719
Root MSE      =  .21667
```

Source	SS	df	MS
Model	50.39547	65	0.77531
Residual	10.93887	233	0.04695
Total	61.33434	298	

	Coef.	Std. Err.	t	P>|t|
Intercept	-0.40061	0.22510	-1.78	0.0764
Age	0.00286	0.01274	0.22	0.8225
707	5.73717	3.57565	1.60	0.1100
720	(dropped)			
727	(omitted)			
737	-0.82298	0.65241	-1.26	0.2084
747	1.24863	0.39107	3.19	0.0016
757	-0.73737	1.25357	-0.59	0.5570
767	0.30373	1.46122	0.21	0.8355
777	(dropped)			
A300	-0.42896	2.10433	-0.20	0.8387
A310	(dropped)			
A319	(dropped)			
A320	(dropped)			
A321	(dropped)			
A330	(dropped)			
BAC111	(dropped)			
CV880	(dropped)			
CV990	(dropped)			
CVR580	-4.28046	3.75698	-1.14	0.2557
DC10	1.06555	0.22470	4.74	0.0000
DC8	0.69845	0.44350	1.57	0.1166
DC9	0.34423	0.25542	1.35	0.1791
F100	(dropped)			
L1011	1.45873	0.47851	3.05	0.0026
L188	(dropped)			
MD11	(dropped)			
MD80	0.17113	0.49750	0.34	0.7312
MD90	(dropped)			
SE210	(dropped)			
V700	(dropped)			
y1965	(dropped)			
y1966	(dropped)			
y1967	(dropped)			
y1968	-0.19914	0.23641	-0.84	0.4005
y1969	(dropped)			
y1970	(dropped)			
y1971	(dropped)			
y1972	(dropped)			
y1973	(dropped)			
y1974	(dropped)			
y1975	(dropped)			
y1976	(dropped)			
y1977	0.15554	0.21398	0.73	0.4680
y1978	0.09983	0.21159	0.47	0.6375
y1979	0.12603	0.19515	0.65	0.5190

y1980	0.23514	0.15712	1.50	0.1359
y1981	0.17959	0.14125	1.27	0.2048
y1982	-0.02521	0.12552	-0.20	0.8410
y1983	-0.14954	0.11644	-1.28	0.2003
y1984	-0.18873	0.10438	-1.81	0.0719
y1985	(dropped)			
y1986	(omitted)			
y1987	-0.01459	0.08648	-0.17	0.8662
y1988	-0.03503	0.09186	-0.38	0.7033
y1989	0.11292	0.09371	1.20	0.2294
y1990	0.15467	0.09464	1.63	0.1035
y1991	0.18693	0.09129	2.05	0.0417
y1992	0.15299	0.09375	1.63	0.1040
y1993	0.07703	0.09827	0.78	0.4339
y1994	0.10218	0.09755	1.05	0.2960
y1995	0.11334	0.09949	1.14	0.2558
y1996	0.16018	0.10231	1.57	0.1188
y1997	0.17973	0.10312	1.74	0.0827
y1998	0.30628	0.10549	2.90	0.0040
y1999	0.29423	0.10817	2.72	0.0070
y2000	0.29086	0.11109	2.62	0.0094
y2001	0.30134	0.11474	2.63	0.0092
y2002	0.19836	0.11790	1.68	0.0938
y2003	0.21364	0.12053	1.77	0.0776
Age*707	-0.43252	0.26554	-1.63	0.1047
Age*720	(dropped)			
Age*727	(omitted)			
Age*737	0.06314	0.04646	1.36	0.1754
Age*747	-0.02508	0.02660	-0.94	0.3466
Age*757	0.08009	0.08682	0.92	0.3573
Age*767	-0.00767	0.10913	-0.07	0.9441
Age*777	(dropped)			
Age*A300	0.08771	0.15939	0.55	0.5827
Age*A310	(dropped)			
Age*A319	(dropped)			
Age*A320	(dropped)			
Age*A321	(dropped)			
Age*A330	(dropped)			
Age*BAC111	(dropped)			
Age*CV880	0.03128	0.01951	1.60	0.1102
Age*CV990	(dropped)			
Age*CVR580	0.13320	0.11755	1.13	0.2584
Age*DC10	-0.01490	0.01159	-1.29	0.1999
Age*DC8	-0.02273	0.02449	-0.93	0.3543
Age*DC9	-0.02308	0.01262	-1.83	0.0686
Age*F100	(dropped)			
Age*L1011	-0.05184	0.02652	-1.95	0.0518
Age*L188	(dropped)			
Age*MD11	(dropped)			
Age*MD80	-0.02300	0.03240	-0.71	0.4784
Age*MD90	(dropped)			
Age*SE210	(dropped)			
Age*V700	(dropped)			
Alaska	(dropped)			
America West	-2.49234	1.07681	-2.31	0.0215
American	(omitted)			

Continental	0.61223	0.35845	1.71	0.0890
Delta	-0.30531	0.24031	-1.27	0.2052
Northwest	-0.28730	0.21562	-1.33	0.1840
Southwest	(dropped)			
United	-0.51138	0.20836	-2.45	0.0148
US Airways	-1.51856	0.60481	-2.51	0.0127
Age*Alaska	0.01197	0.02033	0.59	0.5568
Age*America West	0.15913	0.07358	2.16	0.0316
Age*American	(omitted)			
Age*Continental	-0.01655	0.01650	-1.00	0.3171
Age*Delta	0.01091	0.01325	0.82	0.4111
Age*Northwest	0.00761	0.01161	0.66	0.5130
Age*Southwest	(dropped)			
Age*United	0.03019	0.01209	2.50	0.0132
Age*US Airways	0.10475	0.04018	2.61	0.0097

Short-Lived Fleets Regressions

Table 5.2 presented results suggesting that short-lived fleets do not manifest unusual aging effects. The regressions underlying Table 5.2 are as follows.

Age ≤ 6

```
Number of obs =     355
F( 65,    288) =   12.03
Prob > F      =  0.0000
R-squared     =  0.7308
Adj R-squared =  0.6700
Root MSE      =  .41063
```

Source	SS	df	MS
Model	131.80421	65	2.02776
Residual	48.56238	288	0.16862
Total	180.36659	358	

	Coef.	Std. Err.	t	P>|t|
Intercept	-2.05332	0.24629	-8.34	0.0000
Short Intercept	0.12521	0.17579	0.71	0.4769
Age	0.16814	0.01848	9.10	0.0000
Short Age	0.03463	0.04359	0.79	0.4276
707	-0.44206	0.17935	-2.46	0.0143
720	0.73412	0.24167	-3.04	0.0026
727	(omitted)			
737	0.13715	0.17451	0.79	0.4326

747	1.16801	0.13988	8.35	0.0000
757	0.75587	0.21558	3.51	0.0005
767	0.95380	0.20953	4.55	0.0000
777	1.28197	0.23216	5.52	0.0000
A300	1.25251	0.26101	4.80	0.0000
A310	0.57800	0.31864	1.81	0.0707
A319	0.01805	0.24107	0.07	0.9404
A320	0.56773	0.22354	2.54	0.0116
A321	-0.50619	0.32824	-1.54	0.1241
A330	1.00020	0.31039	3.22	0.0014
BAC111	-0.05293	0.21597	-0.25	0.8066
CV880	-0.38720	0.34786	-1.11	0.2666
CV990	0.11184	0.29157	0.38	0.7016
CVR580	(dropped)			
DC10	0.93902	0.14852	6.32	0.0000
DC8	-0.25316	0.16808	-1.51	0.1331
DC9	-0.21120	0.16495	-1.28	0.2014
F100	0.84980	0.26813	3.17	0.0017
L1011	1.27113	0.17905	7.10	0.0000
L188	(dropped)			
MD11	1.29682	0.24068	5.39	0.0000
MD80	0.10260	0.23641	0.43	0.6646
MD90	0.68790	0.26099	2.64	0.0088
SE210	-0.65851	0.34384	-1.92	0.0565
V700	(dropped)			
y1965	1.63180	0.27801	5.87	0.0000
y1966	1.43110	0.26857	5.33	0.0000
y1967	1.36597	0.27045	5.05	0.0000
y1968	1.29860	0.26161	4.96	0.0000
y1969	1.18483	0.25957	4.56	0.0000
y1970	1.05768	0.25144	4.21	0.0000
y1971	0.83895	0.25228	3.33	0.0010
y1972	0.85580	0.25219	3.39	0.0008
y1973	0.73300	0.25336	2.89	0.0041
y1974	0.85103	0.25737	3.31	0.0011
y1975	0.74222	0.26561	2.79	0.0055
y1976	0.64022	0.26625	2.40	0.0168
y1977	0.60454	0.29049	2.08	0.0383
y1978	0.43200	0.30269	1.43	0.1546
y1979	0.45396	0.36191	1.25	0.2107
y1980	0.36466	0.36244	1.01	0.3152
y1981	0.18007	0.47344	0.38	0.7040
y1982	0.27716	0.33599	0.82	0.4101
y1983	0.03989	0.28885	0.14	0.8903
y1984	0.06835	0.23938	0.29	0.7754
y1985	(dropped)			
y1986	(omitted)			
y1987	0.08562	0.20817	0.41	0.6812
y1988	0.02333	0.21025	0.11	0.9117
y1989	-0.04590	0.20103	-0.23	0.8195
y1990	0.01427	0.20724	0.07	0.9451
y1991	0.18592	0.19638	0.95	0.3446
y1992	-0.00842	0.19538	-0.04	0.9656
y1993	-0.01490	0.19468	-0.08	0.9391
y1994	0.09326	0.21168	0.44	0.6599
y1995	-0.06487	0.21166	-0.31	0.7595

y1996	-0.15804	0.21894	-0.72	0.4710
y1997	-0.08193	0.22621	-0.36	0.7175
y1998	-0.38410	0.22010	-1.75	0.0820
y1999	0.03443	0.20019	0.17	0.8636
y2000	0.13353	0.20338	0.66	0.5120
y2001	0.00646	0.20212	0.03	0.9745
y2002	0.11523	0.20468	0.56	0.5739
y2003	0.15616	0.21127	0.74	0.4604

6 < Age ≤ 12

```
Number of obs =      349
F( 60,    288) =    23.70
Prob > F       =   0.0000
R-squared      =   0.8316
Adj R-squared  =   0.7965
Root MSE       =   .19400
```

Source	SS	df	MS
Model	53.51963	60	.89199
Residual	10.83957	288	.03764
Total	64.35920	348	

	Coef.	Std. Err.	t	P>|t|
Intercept	-1.13550	0.13617	-8.34	0.0000
Short Intercept	0.14985	0.19865	0.75	0.4513
Age	0.03576	0.00846	4.23	0.0000
Short Age	0.00091	0.02139	0.04	0.9661
707	0.10021	0.08705	1.15	0.2506
720	-0.08200	0.12489	-0.66	0.5120
727	(omitted)			
737	0.18632	0.06283	2.97	0.0033
747	1.19487	0.05185	23.04	0.0000
757	0.49649	0.07127	6.97	0.0000
767	0.60158	0.07242	8.31	0.0000
777	(dropped)			
A300	1.20184	0.10422	11.53	0.0000
A310	(dropped)			
A319	(dropped)			
A320	0.31006	0.09132	3.40	0.0008
A321	(dropped)			
A330	(dropped)			
BAC111	(dropped)			
CV880	0.28996	0.14367	2.02	0.0445
CV990	(dropped)			
CVR580	(dropped)			
DC10	0.96198	0.05669	16.97	0.0000
DC8	0.22724	0.07756	2.93	0.0037
DC9	-0.15862	0.08158	-1.94	0.0528

F100	-0.01319	0.15466	-0.09	0.9321
L1011	0.86752	0.08713	9.96	0.0000
L188	-0.03352	0.16282	-0.21	0.8370
MD11	0.83636	0.10408	8.04	0.0000
MD80	0.20546	0.07590	2.71	0.0072
MD90	0.55790	0.13282	4.20	0.0000
SE210	-0.04254	0.15918	-0.27	0.7895
V700	-0.53150	0.23059	-2.31	0.0219
y1965	0.52802	0.24563	2.15	0.0324
y1966	0.55428	0.23512	2.36	0.0191
y1967	0.68298	0.19288	3.54	0.0005
y1968	0.58072	0.18226	3.19	0.0016
y1969	0.52421	0.17619	2.98	0.0032
y1970	0.57838	0.15807	3.66	0.0003
y1971	0.41006	0.15139	2.71	0.0072
y1972	0.28643	0.14079	2.03	0.0428
y1973	0.40223	0.13277	3.03	0.0027
y1974	0.42668	0.12653	3.37	0.0008
y1975	0.37342	0.12528	2.98	0.0031
y1976	0.35459	0.12440	2.85	0.0047
y1977	0.33089	0.12005	2.76	0.0062
y1978	0.24961	0.11876	2.10	0.0364
y1979	0.15083	0.11640	1.30	0.1961
y1980	0.20104	0.11778	1.71	0.0889
y1981	0.17279	0.11755	1.47	0.1427
y1982	0.00011	0.11870	0.00	0.9993
y1983	0.04665	0.11762	0.40	0.6919
y1984	-0.02493	0.12747	-0.20	0.8451
y1985	(dropped)			
y1986	(omitted)			
y1987	0.02104	0.12844	0.16	0.8700
y1988	0.00034	0.12546	0.00	0.9978
y1989	0.09860	0.12686	0.78	0.4377
y1990	0.14058	0.12483	1.13	0.2610
y1991	0.13797	0.14091	0.98	0.3283
y1992	-0.07576	0.14899	-0.51	0.6115
y1993	-0.07934	0.14902	-0.53	0.5948
y1994	-0.15595	0.13613	-1.15	0.2529
y1995	-0.14954	0.13311	-1.12	0.2622
y1996	-0.15650	0.13061	-1.20	0.2318
y1997	-0.10245	0.13083	-0.78	0.4342
y1998	-0.09289	0.11896	-0.78	0.4355
y1999	-0.05888	0.11920	-0.49	0.6217
y2000	0.06541	0.11875	0.55	0.5822
y2001	0.21424	0.11892	1.80	0.0727
y2002	0.19402	0.11970	1.62	0.1061
y2003	0.05019	0.12145	0.41	0.6798

12 < Age

```
Number of obs =      299
F( 42,    256) =   17.85
Prob > F       =  0.0000
```

```
R-squared     =  0.7455
Adj R-squared =  0.7037
Root MSE      =  .24696

      Source |       SS         df         MS
-------------+------------------------------
       Model |  45.72172        42     1.08861
    Residual |  15.61262       256     0.06099
-------------+------------------------------
       Total |  61.33434       298

-------------------------------------------------------------
             |      Coef.    Std. Err.        t     P>|t|
-------------+-----------------------------------------------
   Intercept | -0.50374     0.13021       -3.87    0.0001
Short Intercept |  0.17352  0.45896        0.38    0.7057
         Age |  0.00676     0.00742        0.91    0.3635
   Short Age | -0.00937     0.03022       -0.31    0.7567
         707 | -0.06484     0.19752       -0.33    0.7430
         720 |  (dropped)
         727 |  (omitted)
         737 |  0.03447     0.06820        0.51    0.6137
         747 |  0.81824     0.06267       13.06    0.0000
         757 |  0.35275     0.12082        2.92    0.0038
         767 |  0.19136     0.13549        1.41    0.1591
         777 |  (dropped)
        A300 |  0.76563     0.16802        4.56    0.0000
        A310 |  (dropped)
        A319 |  (dropped)
        A320 |  (dropped)
        A321 |  (dropped)
        A330 |  (dropped)
      BAC111 |  (dropped)
       CV880 |  0.21013     0.28424        0.74    0.4604
       CV990 |  (dropped)
      CVR580 | -0.08454     0.19851       -0.43    0.6706
        DC10 |  0.79611     0.04615       17.25    0.0000
         DC8 |  0.27814     0.07759        3.58    0.0004
         DC9 | -0.24459     0.06590       -3.71    0.0003
        F100 |  (dropped)
       L1011 |  0.49229     0.08234        5.98    0.0000
        L188 |  (dropped)
        MD11 |  (dropped)
        MD80 | -0.02282     0.09600       -0.24    0.8123
        MD90 |  (dropped)
       SE210 |  (dropped)
        V700 |  (dropped)
       y1965 |  (dropped)
       y1966 |  (dropped)
       y1967 |  (dropped)
       y1968 | -0.34793     0.28560       -1.22    0.2243
       y1969 |  (dropped)
       y1970 |  (dropped)
       y1971 |  (dropped)
       y1972 |  (dropped)
       y1973 |  (dropped)
```

```
     y1974 |  (dropped)
     y1975 |  (dropped)
     y1976 |  (dropped)
     y1977 |  0.19682      0.20044        0.98    0.3271
     y1978 |  0.13351      0.20022        0.67    0.5055
     y1979 |  0.23980      0.17850        1.34    0.1803
     y1980 |  0.22999      0.15405        1.49    0.1367
     y1981 |  0.15412      0.13993        1.10    0.2717
     y1982 |  0.04425      0.13588        0.33    0.7450
     y1983 | -0.07892      0.12652       -0.62    0.5333
     y1984 | -0.13515      0.11442       -1.18    0.2386
     y1985 |  (dropped)
     y1986 |  (omitted)
     y1987 | -0.01385      0.09744       -0.14    0.8871
     y1988 | -0.01484      0.09962       -0.15    0.8817
     y1989 |  0.11662      0.10483        1.11    0.2670
     y1990 |  0.15557      0.10563        1.47    0.1420
     y1991 |  0.20231      0.10108        2.00    0.0464
     y1992 |  0.15578      0.10343        1.51    0.1333
     y1993 |  0.06304      0.10822        0.58    0.5607
     y1994 |  0.04794      0.10551        0.45    0.6500
     y1995 |  0.06041      0.10714        0.56    0.5733
     y1996 |  0.10512      0.10921        0.96    0.3367
     y1997 |  0.11507      0.10857        1.06    0.2902
     y1998 |  0.27686      0.10648        2.60    0.0099
     y1999 |  0.24316      0.10884        2.23    0.0263
     y2000 |  0.27394      0.11261        2.43    0.0157
     y2001 |  0.29729      0.11425        2.60    0.0098
     y2002 |  0.19529      0.12012        1.63    0.1052
     y2003 |  0.24537      0.12137        2.02    0.0442
------------------------------------------------------
```

Finally, Table 5.3 suggested that short-lived fleets are unusually expensive when they are new and when they are in the mature period. The follow regressions underlie Table 5.3.

Age ≤ 6

```
Number of obs =     355
F( 64,   289) =   12.22
Prob > F      =  0.0000
R-squared     =  0.7302
Adj R-squared =  0.6704
Root MSE      =  .41037

      Source |       SS        df        MS
-------------+------------------------------
       Model |  131.69777      64     2.05778
    Residual |   48.66882     289     0.16840
-------------+------------------------------
       Total |  180.36659     353
------------------------------------------------------
```

	Coef.	Std. Err.	t	P>|t|
Intercept	-2.05042	0.24610	-8.33	0.0000
Short Intercept	0.23489	0.10876	2.16	0.0316
Age	0.17283	0.01750	9.87	0.0000
707	-0.42658	0.17817	-2.39	0.0173
720	-0.66940	0.22738	-2.94	0.0035
727	(omitted)			
737	0.13532	0.17438	0.78	0.4384
747	1.16164	0.13956	8.82	0.0000
757	0.75492	0.21544	3.50	0.0005
767	0.94824	0.20928	4.53	0.0000
777	1.28525	0.23198	5.54	0.0000
A300	1.25573	0.26081	4.81	0.0000
A310	0.62826	0.31210	2.01	0.0450
A319	0.02398	0.24080	0.10	0.9207
A320	0.56770	0.22340	2.54	0.0116
A321	-0.49903	0.32791	-1.52	0.1291
A330	1.01020	0.30994	3.26	0.0012
BAC111	-0.05680	0.21577	-0.26	0.7925
CV880	-0.32127	0.33760	-0.95	0.3421
CV990	0.15848	0.28542	0.56	0.5792
CVR580	(dropped)			
DC10	0.93035	0.14802	6.29	0.0000
DC8	-0.25662	0.16791	-1.53	0.1275
DC9	-0.20948	0.16483	-1.27	0.2048
F100	0.85544	0.26786	3.19	0.0016
L1011	1.27288	0.17892	7.11	0.0000
L188	(dropped)			
MD11	1.29613	0.24053	5.39	0.0000
MD80	0.11014	0.23607	0.47	0.6412
MD90	0.68882	0.26082	2.64	0.0087
SE210	-0.60336	0.33655	-1.79	0.0740
V700	(dropped)			
y1965	1.60153	0.27521	5.82	0.0000
y1966	1.40919	0.26698	5.28	0.0000
y1967	1.34415	0.26889	5.00	0.0000
y1968	1.27846	0.26022	4.91	0.0000
y1969	1.17090	0.25881	4.52	0.0000
y1970	1.04684	0.25091	4.17	0.0000
y1971	0.82871	0.25179	3.29	0.0011
y1972	0.83737	0.25096	3.34	0.0010
y1973	0.71503	0.25219	2.84	0.0049
y1974	0.84199	0.25696	3.28	0.0012
y1975	0.73691	0.26536	2.78	0.0058
y1976	0.63303	0.26592	2.38	0.0179
y1977	0.58638	0.28940	2.03	0.0437
y1978	0.41084	0.30132	1.36	0.1738
y1979	0.42953	0.36037	1.19	0.2343
y1980	0.33792	0.36065	0.94	0.3496
y1981	0.15193	0.47181	0.32	0.7477
y1982	0.26302	0.33531	0.78	0.4334
y1983	0.03880	0.28866	0.13	0.8932
y1984	0.05799	0.23887	0.24	0.8084
y1985	(dropped)			
y1986	(omitted)			

y1987	0.07859	0.20785	0.38	0.7056
y1988	0.00867	0.20930	0.04	0.9670
y1989	-0.05892	0.20023	-0.29	0.7688
y1990	0.00007	0.20633	0.00	0.9997
y1991	0.16419	0.19435	0.84	0.3989
y1992	-0.02781	0.19373	-0.14	0.8859
y1993	-0.03963	0.19206	-0.21	0.8367
y1994	0.07009	0.20953	0.33	0.7382
y1995	-0.08198	0.21043	-0.39	0.6971
y1996	-0.17314	0.21797	-0.79	0.4277
y1997	-0.09608	0.22536	-0.43	0.6702
y1998	-0.39198	0.21973	-1.78	0.0755
y1999	0.01770	0.19895	0.09	0.9292
y2000	0.11856	0.20238	0.59	0.5584
y2001	-0.00944	0.20100	-0.05	0.9626
y2002	0.09569	0.20307	0.47	0.6378
y2003	0.13330	0.20917	0.64	0.5245

6 < Age ≤ 12

```
Number of obs =      349
F( 59,    289) =   24.18
Prob > F       =  0.0000
R-squared      =  0.8316
Adj R-squared  =  0.7972
Root MSE       =  .19367
```

Source	SS	df	MS
Model	53.51956	59	0.90711
Residual	10.83964	289	0.03751
Total	64.35920	348	

	Coef.	Std. Err.	t	P>|t|
Intercept	-1.13667	0.13314	-8.54	0.0000
Short Intercept	0.15799	0.05424	2.91	0.0039
Age	0.03588	0.00793	4.52	0.0000
707	0.09998	0.08674	1.15	0.2500
720	-0.08064	0.12053	-0.67	0.5040
727	(omitted)			
737	0.18606	0.06243	2.98	0.0031
747	1.19484	0.05176	23.09	0.0000
757	0.49616	0.07071	7.02	0.0000
767	0.60123	0.07182	8.37	0.0000
777	(dropped)			
A300	1.20152	0.10377	11.58	0.0000
A310	(dropped)			
A319	(dropped)			
A320	0.30997	0.09114	3.40	0.0008
A321	(dropped)			

```
    A330 |  (dropped)
  BAC111 |  (dropped)
   CV880 |   0.29152      0.13865       2.10   0.0364
   CV990 |  (dropped)
  CVR580 |  (dropped)
    DC10 |   0.96208      0.05655      17.01   0.0000
     DC8 |   0.22743      0.07730       2.94   0.0035
     DC9 |  -0.15874      0.08138      -1.95   0.0521
    F100 |  -0.01363      0.15405      -0.09   0.9296
   L1011 |   0.86729      0.08680       9.99   0.0000
    L188 |  -0.03103      0.15168      -0.20   0.8380
    MD11 |   0.83578      0.10300       8.11   0.0000
    MD80 |   0.20506      0.07519       2.73   0.0068
    MD90 |   0.55785      0.13259       4.21   0.0000
   SE210 |  -0.04149      0.15699      -0.26   0.7917
    V700 |  -0.52662      0.19971      -2.64   0.0088
   y1965 |   0.52337      0.21955       2.38   0.0178
   y1966 |   0.55041      0.21647       2.54   0.0115
   y1967 |   0.67990      0.17848       3.81   0.0002
   y1968 |   0.57830      0.17290       3.34   0.0009
   y1969 |   0.52256      0.17160       3.05   0.0025
   y1970 |   0.57778      0.15718       3.68   0.0003
   y1971 |   0.40975      0.15095       2.71   0.0070
   y1972 |   0.28618      0.14043       2.04   0.0425
   y1973 |   0.40199      0.13242       3.04   0.0026
   y1974 |   0.42672      0.12631       3.38   0.0008
   y1975 |   0.37352      0.12504       2.99   0.0031
   y1976 |   0.35478      0.12411       2.86   0.0046
   y1977 |   0.33122      0.11960       2.77   0.0060
   y1978 |   0.24996      0.11827       2.11   0.0354
   y1979 |   0.15101      0.11612       1.30   0.1945
   y1980 |   0.20126      0.11747       1.71   0.0877
   y1981 |   0.17307      0.11715       1.48   0.1407
   y1982 |   0.00010      0.11850       0.00   0.9993
   y1983 |   0.04665      0.11742       0.40   0.6915
   y1984 |  -0.02505      0.12722      -0.20   0.8440
   y1985 |  (dropped)
   y1986 |  (omitted)
   y1987 |   0.02076      0.12804       0.16   0.8713
   y1988 |  -0.00003      0.12494      -0.00   0.9998
   y1989 |   0.09875      0.12659       0.78   0.4360
   y1990 |   0.14096      0.12430       1.13   0.2577
   y1991 |   0.13891      0.13895       1.00   0.3183
   y1992 |  -0.07496      0.14754      -0.51   0.6118
   y1993 |  -0.07842      0.14717      -0.53   0.5946
   y1994 |  -0.15534      0.13513      -1.15   0.2513
   y1995 |  -0.14900      0.13228      -1.13   0.2609
   y1996 |  -0.15600      0.12986      -1.20   0.2306
   y1997 |  -0.10199      0.13016      -0.78   0.4339
   y1998 |  -0.09250      0.11841      -0.78   0.4353
   y1999 |  -0.05859      0.11880      -0.49   0.6223
   y2000 |   0.06574      0.11829       0.56   0.5788
   y2001 |   0.21461      0.11839       1.81   0.0709
   y2002 |   0.19433      0.11927       1.63   0.1043
   y2003 |   0.05047      0.12105       0.42   0.6770
-------------------------------------------------------
```

12 < Age

```
Number of obs =     299
F( 41,   257) =   18.35
Prob > F      =  0.0000
R-squared     =  0.7454
Adj R-squared =  0.7047
Root MSE      =  .24652
```

Source	SS	df	MS
Model	45.71586	41	1.11502
Residual	15.61849	257	0.06077
Total	61.33434	298	

	Coef.	Std. Err.	t	P>|t|
Intercept	-0.49979	0.12935	-3.86	0.0001
Short Intercept	0.03360	0.08407	0.40	0.6897
Age	0.00649	0.00736	0.88	0.3789
707	-0.05103	0.19210	-0.27	0.7907
720	(dropped)			
727	(omitted)			
737	0.03283	0.06788	0.48	0.6290
747	0.81885	0.06253	13.10	0.0000
757	0.35055	0.12040	2.91	0.0039
767	0.18806	0.13483	1.39	0.1643
777	(dropped)			
A300	0.76268	0.16746	4.55	0.0000
A310	(dropped)			
A319	(dropped)			
A320	(dropped)			
A321	(dropped)			
A330	(dropped)			
BAC111	(dropped)			
CV880	0.23423	0.27293	0.86	0.3916
CV990	(dropped)			
CVR580	-0.08026	0.19769	-0.41	0.6851
DC10	0.79661	0.04604	17.30	0.0000
DC8	0.27763	0.07743	3.59	0.0004
DC9	-0.24357	0.06570	-3.71	0.0003
F100	(dropped)			
L1011	0.49189	0.08218	5.99	0.0000
L188	(dropped)			
MD11	(dropped)			
MD80	-0.02982	0.09314	-0.32	0.7491
MD90	(dropped)			
SE210	(dropped)			
V700	(dropped)			
y1965	(dropped)			
y1966	(dropped)			
y1967	(dropped)			
y1968	-0.32246	0.27306	-1.18	0.2387

```
y1969 |  (dropped)
y1970 |  (dropped)
y1971 |  (dropped)
y1972 |  (dropped)
y1973 |  (dropped)
y1974 |  (dropped)
y1975 |  (dropped)
y1976 |  (dropped)
y1977 |  0.19673      0.20009       0.98    0.3264
y1978 |  0.13347      0.19987       0.67    0.5049
y1979 |  0.24188      0.17806       1.36    0.1755
y1980 |  0.23084      0.15376       1.50    0.1345
y1981 |  0.15295      0.13963       1.10    0.2744
y1982 |  0.04959      0.13455       0.37    0.7127
y1983 | -0.07607      0.12596      -0.60    0.5464
y1984 | -0.13435      0.11419      -1.18    0.2405
y1985 |  (dropped)
y1986 |  (omitted)
y1987 | -0.01365      0.09727      -0.14    0.8885
y1988 | -0.01378      0.09938      -0.14    0.8898
y1989 |  0.11686      0.10464       1.12    0.2651
y1990 |  0.15602      0.10544       1.48    0.1402
y1991 |  0.20285      0.10089       2.01    0.0454
y1992 |  0.15766      0.10307       1.53    0.1274
y1993 |  0.06346      0.10802       0.59    0.5574
y1994 |  0.05079      0.10492       0.48    0.6288
y1995 |  0.06275      0.10669       0.59    0.5569
y1996 |  0.10696      0.10886       0.98    0.3267
y1997 |  0.11631      0.10830       1.07    0.2838
y1998 |  0.27756      0.10627       2.61    0.0095
y1999 |  0.24319      0.10865       2.24    0.0261
y2000 |  0.27463      0.11239       2.44    0.0152
y2001 |  0.29824      0.11401       2.62    0.0094
y2002 |  0.19850      0.11946       1.66    0.0978
y2003 |  0.03360      0.12065       2.06    0.0402
------------------------------------------------
```

Bibliography

Aviation Safety Alliance, "Aviation 101: Maintenance" no date (http://www.aviationsafetyalliance.org/aviation/maintenance.asp; accessed December 28, 2005).

Boeing, "Airframe Maintenance Cost Analysis Methodology," briefing presented to RAND Corporation, September 2, 2004a.

Boeing, *The Boeing Company 2003 Annual Report,* "Notes to Consolidated Financial Statements," 2004b (http://www.boeing.com/company offices/financial/finreports/annual/03annualreport/f_ncfs_08.html; accessed January 3, 2006).

Boeing, "Commercial Airplanes—Jetliner Safety: Airline's Role in Aviation Safety, 1995–2006," no date (http://www.boeing.com/commercial/safety/airline_role.html; accessed December 28, 2005.

"Boeing Wins Modification and Maintenance Contract with Kitty Hawk International for Five 747 Airplanes," Boeing News release, June 4, 1999 (http://www.boeing.com/news/releases/1999/news_release_990604a.html; accessed December 28, 2005).

Bolkcom, Christopher, *The Air Force KC-767 Tanker Lease Proposal: Key Issues For Congress,* Congressional Research Service Report for Congress, September 2, 2003.

Bureau of Economic Analysis, *National Economic Accounts,* "Implicit Price Deflators for Gross Domestic Product," Table 1.1.9, December 2004 (http://www.bea.gov/bea/dn/nipaweb/SelectTable.asp?Selected=N; accessed March 9, 2006).

Bureau of Transportation Statistics, *TranStats,* Form 41 data, January 3, 2006 (http://www.transtats.bts.gov/Databases.asp?Mode_ID=1&Mode_Desc=Aviation&Subject_ID2=0; accessed January 3, 2006).

Francis, Peter J., and Geoffrey B. Shaw, *Effect of Aircraft Age on Maintenance Costs,* Alexandria, Va.: Center for Naval Analyses, CAB D0000289.A2, March 2000.

General Accounting Office (now the Government Accountability Office), *Aging Refueling Aircraft Are Costly to Maintain and Operate,* Report to Congressional Committees, Washington, D.C.: GAO, August 1996.

Greenfield, Victoria A., and David M. Persselin, *An Economic Framework for Evaluating Military Aircraft Replacement,* Santa Monica, Calif.: RAND Corporation, MR-1489-AF, 2002.

Hildebrandt, Gregory G., and Man-bing Sze, *An Estimation of USAF Aircraft Operating and Support Cost Relations,* Santa Monica, Calif.: RAND Corporation, N-3062-ACQ, May 1990.

Johnson, John, *Age Impacts on Operating and Support Costs: Navy Aircraft Age Analysis Methodology,* Patuxent River, Md.: Naval Aviation Maintenance Office, August 1993.

Jondrow, James, Robert P. Trost, Michael Ye, John P. Hall, Rebecca L. Kirk, Laura J. Junor, Peter J. Francis, Geoffrey B. Shaw, Darlene E. Stanford, and Barbara H. Measell, *Support Costs and Aging Aircraft: Implications for Budgeting and Procurement,* Alexandria, Va.: CNA Corporation, January 2002.

Kamins, Milton, *The Effect of Calendar Age on Aircraft Maintenance Requirements,* Santa Monica, Calif.: RAND Corporation, WN-7167-PR, December 1970.

Keating, Edward G., and Matthew Dixon, *Investigating Optimal Replacement of Aging Air Force Systems,* Santa Monica, Calif.: RAND Corporation, MR-1763-AF, 2003.

Kiley, Gregory T., *The Effects of Aging on the Costs of Maintaining Military Equipment,* Washington, D.C.: Congressional Budget Office, August 2001.

"Maintenance: Turning the Screw," *Airline Business,* September 26, 2005 (http://www.flightinternational.com/Articles/2005/09/26/Navigation/199/201685/Maintenance+Turning+the+screw.html; accessed January 3, 2006).

Peters, John E., and Benjamin Zycher, *Analytics of Third-Party Claim Recovery for Military Aircraft Engine Warranties,* Santa Monica, Calif.: RAND Corporation, DB-368-OSD, 2002.

Pyles, Raymond A., *Aging Aircraft: USAF Workload and Material Consumption Life Cycle Patterns,* Santa Monica, Calif.: RAND Corporation, MR-1641-AF, 2003.

Ramsey, Tom, Carl French, and Kenneth R. Sperry, "Airframe Maintenance Trend Analysis," briefing, Oklahoma City ALC (Ramsey) and Boeing (French and Sperry), 1998.

Stoll, Lawrence, and Stan Davis, *Aircraft Age Impact on Individual Operating and Support Cost Elements,* Patuxent River, Md.: Naval Aviation Maintenance Office, July 1993.

Wikipedia, "Aircraft maintenance checks," no date (http://en.wikipedia.org/wiki/Aircraft_maintenance_checks, accessed December 28, 2005).

Witkin, Richard, "Problems Detected in Engine Mounting on 37 DC-10 Planes," *The New York Times,* June 1, 1979, pp. A1, A10.